Fly Ash and Coal Conversion By-Products: Characterization, Utilization, and Disposal I

MATERIALS RESEARCH SOCIETY SYMPOSIA PROCEEDINGS

ISSN 0272 - 9172

MATERIALS RESEARCH SOCIETY SYMPOSIA PROCEEDINGS

MATERIALS RESEARCH SOCIETY SYMPOSIA PROCEEDINGS   VOLUME 43

Fly Ash and Coal Conversion By-Products Characterization, Utilization, and Disposal

Symposium held November 29-30, 1984, Boston, Massachusetts, U. S. A.

EDITORS:

Gregory J. McCarthy

North Dakota State University, Fargo, North Dakota, U. S. A.

Robert J. Lauf

Oak Ridge National Laboratory, Oak Ridge, Tennessee, U. S. A.

MRS   MATERIALS RESEARCH SOCIETY
Pittsburgh, Pennsylvania

CAMBRIDGE UNIVERSITY PRESS
Cambridge, New York, Melbourne, Madrid, Cape Town,
Singapore, São Paulo, Delhi, Mexico City

Cambridge University Press
32 Avenue of the Americas, New York NY 10013-2473, USA

Published in the United States of America by Cambridge University Press, New York

www.cambridge.org
Information on this title: www.cambridge.org/9781107405677

Materials Research Society
506 Keystone Drive, Warrendale, PA 15086
http://www.mrs.org

First published 1985
First paperback edition 2012

Single article reprints from this publication are available through
University Microfilms Inc., 300 North Zeeb Road, Ann Arbor, MI 48106

CODEN: MRSPDH

ISBN 978-0-931-83708-1 Hardback
ISBN 978-1-107-40567-7 Paperback

Contents

Preface

Vast quantities of inorganic by-products are produced when coal is burned or gasified. In the United States, nearly 60 million tons of fly ash is removed from power plant stacks each year and more than 80% of this ash is buried in landfills or stored in holding ponds pending burial. Only 10% has found a commercial market, chiefly as in concrete-related uses. Conversion of coal to other energy sources will lead to additional large quantities of ash in the future. In 1984, the first commercial US coal gasification plant opened in North Dakota. The 1100 tons of ash produced each day is being buried in the nearby strip mine. With such vast quantities of ash involved, one needs to be attuned to any environmental consequences of burying ash and to any new possibilities of using this vast resource for industrial or civil engineering applications. Thorough characterization of the ash materials and their reactions provides the scientific basis of safe disposal or effective utilization. This volume brings together peer-reviewed papers on characterization, utilization and disposal of coal fly ash and gasification ash.

The papers in this volume are based on presentations in Symposium M of the 1984 Fall Meeting of the Materials Research Society. This was the the Society's third symposium on the subject of ash characterization and utilization. Proceedings of the first were published under the title "Effects of Fly Ash Incorporation in Cement and Concrete," edited by Sidney Diamond. Copies of these proceedings are available from D. M. Roy of Penn State's Materials Research Laboratory. Relevant papers from the second symposium, edited by G. J. McCarthy, appeared in the July 1984 issue of Cement and Concrete Research. The Society will hold a fourth symposium on this subject in 1985 and the proceedings will be published as Volume II under the title of this book.

The state of the art in the characterization of fly ash and fly ash-containing mortars and concrete is the subject of the keynote paper prepared by Professor Della M. Roy of Penn State University. X-ray diffraction, scanning electron microscopy and thermal analysis are most often employed in characterizing these materials and their reactions. These techniques are highlighted in many of the papers that follow, along with other methods ranging from time-honored petrographic microscopy and metallography to the first results to be obtained from the new microfocus laser Raman instruments.

Utilization research presented here includes the well-established replacement of some of the cement in concrete by fly ash as well as roadbed stabilization and oil and gas well cementing applications. Characterization and potential uses for gasification ash are presented by scientists and engineers from the North Dakota universities. Processes that employ coal ash as a source of aluminum are discussed in a paper from the Oak Ridge National Laboratory. While not economical at present, coal ash could become a principal Al source in case of any future bauxite shortage.

Environmental concerns in use of fly ash for roadbed stabilization and in burial of fly ash in reclaimed strip mines and municipal landfills are addressed in the final section of this volume. The formation from coal waste of an artificial reef described by Hornibrook and Parker is an innovative combination of utilization and disposal.

It is hoped that this volume, combined with the previous two proceedings, will provide the interested reader with up-to-date information on coal combustion and conversion by-products with a distinct materials science and engineering flavor.

R. J. Lauf and I would like to acknowledge the financial and logistical support provided by the Chemistry Department at North Dakota State University and by Oak Ridge National Laboratory in preparing this volume. In organizing the Symposium we had the invaluable assistance of Professors Turgut Demirel of Iowa State and Barry E. Scheetz of Penn State. We are also grateful to the Tennessee Valley Authority for a grant that helped to make publication of these proceedings possible.

G. J. McCarthy
Fargo, North Dakota
February, 1985

PART I

Fly Ash Characterization

CHARACTERIZATION OF FLY ASH AND ITS REACTIONS IN CONCRETE

DELLA M. ROY*, KAREN LUKE**, AND SIDNEY DIAMOND***
*Materials Research Laboratory, The Pennsylvania State University, University Park, PA 16802, USA
**Department of Chemistry, University of Aberdeen, Old Aberdeen AB9 2UE, Scotland
***School of Civil Engineering, Purdue University, Lafayette, IN 47906, USA

(Received 12 February, 1985; Communicated by G.J. McCarthy)

ABSTRACT

Fly ashes are currently being produced that are much more widely different from each other in composition and other characteristics than had been previously experienced, owing to the widespread use of low rank sub-bituminous and lignitic coals. The current ASTM classifications into Class F and Class C pozzolan categories are not adequate to describe all their important properties. Current characterization methods are reviewed, including physical characterization by particle size distribution, shape, apparent specific gravity, content of hollow grains and of residual coal fragments, etc., chemical procedures of various kinds, and SEM, EDXA, XRD, and other methods for the determination of mineralogical content and glass character. Etching and chemical dissolution procedures are particularly important. The state of these various methods, current results of their use in rly ash characterizations, and the relations of these to reactivity and performance of fly ashes in cement and concrete are discussed.

INTRODUCTION

Fly ash is a material whose characteristics reflect its origin from incombustible mineral matter in powdered coal burned in large electrical power plant boilers. Because of extensive variability both in the coals and in the operation of the boilers and collection of the fly ash, fly ashes differ from each other in important characteristics. Moreover, the chemical composition and other characteristics of different fly ash particles in a given ash may be extremely different from each other, reflecting the particular bits of mineral matter giving rise to that particle. These complications render the appropriate characterization of fly ashes for use in concrete and elsewhere difficult and intricate [1a].

By definition, all particles of fly ash are 'fine'--at least fine enough to have been swept along in the air stream and delivered to the electrostatic precipitator or baghouse collection system. Actual sizes may vary from several hundred micrometers to several tenths of a micrometer.

The particles are primarily spherical, although incompletely rounded spheres, rough, vacuole-containing massy agglomerates, and unburned coal fragments are also found. The spheres are primarily solid, although hollow spheres occur to a small extent in many fly ashes. The spheres are mostly glassy, although crystals are usually found within them, and deposits of crystalline powders of several kinds on the spherical surfaces are not infrequently found.

The relative proportions of spherical and other particles, the size distributions, the kind of crystals present, the nature of the glass, the type of surface deposit, if any, the nature and proportion of unburned coal, and many other parameters vary among different fly ashes. These variations, and accompanying variations in overall chemical composition result in variations in behavior when fly ash is used in concrete, and in other uses as well.

Use of fly ash as a cementitious substitute or mineral admixture in concrete is a major application of the material, although only about 6% of the total produced in the U.S. is currently used in the concrete industry [1b]. A major factor limiting more extensive use is the intrinsic variability of the material, coupled with the complexity of its characterization. Characterization and specification methods currently prescribed are generally not adequate to insure optimum selection of fly ash for use in concrete, nor to insure adequate quality control on a continuing basis.

The purpose of the present paper is to review the current status of fly ash characterization, especially with regard to parameters that may influence behavior and suitability of fly ash for use in concrete, specifically:

> Methodologies for characterization,
> Morphology, composition, and mineralogy,
> Chemical composition and physical properties,
> Classification, and
> Reactions with cement and concrete components.

Table I. Methodology for Characterization.

METHODOLOGIES FOR CHARACTERIZATION

Chemical
Chemical Analysis – Bulk (including LOI)
Available Alkalies

Physical
Pozzolanic Reactivity – Lime (7 days)
                      Cement (28 days)
Specific Gravity
Autoclave Expansion
Fineness   Sieve %
              Blaine
              BET (N_2)
              Sedigraph (Liquid), Coulter
              SEM

Other
Mineralogical Analysis – (Qualitative, Quantitative)
   X-ray Diffraction Phase Analysis
   Microscopic (Optical and Scanning [SEM])
   TEM (rarely)
IR, Raman Spectroscopy
SEM, EDS – Morphology/Chemistry
Selective Dissolution
Soluble Components

Insight into the behavior of fly ash in concrete can be achieved most suitably by a broad, multi-method approach to characterization, although not all methods are equally valuable. Table I contains a list of methodologies which may be important. The list includes the chemical and physical procedures specified in ASTM C 618 and C 311, but is certainly not limited to these relatively simple standard procedures. It is commonly found that the results of the standard procedures are by themselves inadequate to explain behavior, and in view of the complexity and variability of fly ashes this is not surprising. It is thus necessary to consider a significantly longer list of methodologies which may serve in due course as a source for upgrading standard procedures. The state of the art at this stage is probably not yet ready for the proposal of a new comprehensive characterization scheme, but this paper, along with others in the symposium, may help to provide the needed input.

Morphology, Composition, and Mineralogy

Spherical fly ash particles get their shape as a result of cooling and solidifying from molten droplets of inorganic coal residue. The surfaces are often quite smooth and 'glassy' in character, but in some ashes they tend to be rough and pebbly on a micro scale, and in others may be partly covered with a deposit of powder condensed from the vapor phase after the spheres have solidified (Fig. 1). The material deposited is often alkali sulfate [1] and as such it is readily soluble.

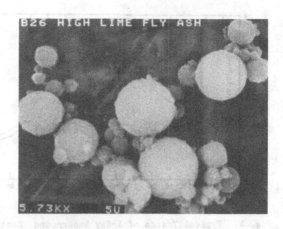

Fig. 1. High-Calcium Fly Ash Spherical Particles, Showing Adherence of Finer Particles to the Surface (bar = 5µm) [2].

Some particles solidify around a trapped gas bubble and are thus hollow spheres, frequently thin-walled. Additionally, hollow spheres containing smaller incorporated fly ash spheres are found on occasion, with the incorporated spheres being mostly solid, but themselves occasionally hollow. Curiously, a third level of incorporated spheres can sometimes be seen within hollow spheres incorporated in larger hollow spheres, but these are rare.

Empty hollow spheres are usually called cenospheres; those containing included spheres, plerospheres. Factors involved in their formation are not as yet fully understood. Cenospheres can be readily segregated from other fly ash particles by flotation methods, and are used in several industrial applications.

Cenospheres and plerospheres are often recognized in scanning electron microscope (SEM) study of fly ashes by the fact that shells may be incomplete or broken in spots. Completely unbroken hollow spheres cannot be recognized as such, but may be detected in optical microscopy or if cut through in special mounting or sectioning procedures.

The hollow spheres are ordinarily completely glassy; solid spheres may be so, but usually contain incorporated crystals of various kinds embedded within the glass matrix.

The nature of the glass present in fly ashes is coming under serious study, since glass is usually the predominating phase, constituting about 60 to 90 percent of most fly ashes. It has been determined [1-7] that differences exist in the structure of the glass in low-calcium ('Class F') and in high-calcium ('Class C') fly ashes. Glasses give rise to an amorphous scattering 'hump' in x-ray diffraction, and the differences are reflected in the position of the maximum of this broad hump (Fig. 2). The glass structure of the low calcium fly ashes is of a siliceous type [2,5,6], and it is likely that the structure, suitably modified, can accommodate increasing CaO contents found in high-calcium fly ashes of up to perhaps 20% total CaO content (a significant proportion of which is incorporated in

6

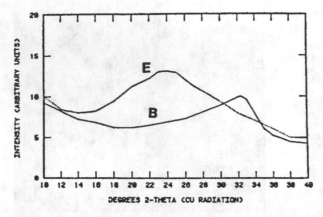

Fig. 2. Typical Traces of X-Ray Background Scattering 'Humps' for Fly Ash B (High Calcium) and Fly Ash E (Low Calcium), with the Crystalline Phase Subtracted [4].

crystalline components in most such fly ashes). Fly ashes with overall CaO contents above 20% likely will form glass with a calcium aluminate type structure, a melilite (= pyrosilicate, Si_2O_7 or $SiAlO_7$ structure) type silicate glass, or sometimes a mixture of glass types. Calcium aluminate glasses are believed to be more reactive in concrete.

Table II. Common Crystalline Phases in Class F Fly Ashes.

mullite	$Al_6Si_2O_{13}$
quartz	SiO_2
magnetite-ferrite	Fe_3O_4-$(Mg,Fe)(Fe,Mg)_2O_4$
hematite	Fe_2O_3
anhydrite	$CaSO_4$

In addition to the glass, there are a large number of crystalline phases present in most fly ashes. With few exceptions the individual crystalline phases occur only in small amounts, usually only a few percent at most of each type being present. As indicated in Table II, the common suite found in most low-calcium fly ashes and others as well includes —quartz, mullite, hematite, and magnetite (although the last-named phase is better described as a spinel). Anhydrite is found in some higher CaO content Class F fly ashes, and in many Class C fly ashes as well.

Many fly ashes contain alkali sulfates, ordinarily as surface deposits. The crystalline species present include potassium sulfate (α-K_2SO_4), sodium sulfate (thenardite), and mixed sulfates, principally aphthitalite [$(Na,K)_2SO_4$] and potassium-calcium sulfates.

A further group of crystalline components are calcium compounds primarily associated with high-calcium fly ashes, as listed in Table III [8] and Fig. 3. Several of these can become involved in various hydration reactions, so that they are of particular interest in concrete applications.

In many of these fly ashes a portion of the overall analytical CaO content is present as crystalline CaO, typically amounting to between 2% and 5% CaO. Individual crystalline particles are not ordinarily detected, and the crystalline material is either embedded within the glass or deposited as a film on its surface [7–13]. From its ready availability to solution, the latter seems more likely. The dissolved CaO is capable of reacting with the glass phase in a 'self-pozzolanic' reaction, but this is relatively slow. When crystalline CaO is not reactive upon first contact with solutions, it can remain to cause unsoundness problems because of the reactive lime to portlandite reaction [14].

Table III.   Class C Fly Ash Primary Crystalline Phases.

melilite (gehlenite-akermanite)	$Ca, Mg, Al(Si_2O_7)$
ferrite spinel	$(Mg, Fe)(Fe, Mg)_2O_4$
merwinite	$Ca_3Mg(SiO_4)_2$
bredigite, larnite	Ca_2SiO_4
lime	CaO
periclase	MgO
nepheline, carnegieite	$NaAlSiO_4$
feldspar	$(Na, Ca, Al, Silicate)$
pyroxene	$(Mg, Fe, Ca, Al, Silicate)$
cristobalite/quartz	SiO_2
C_3S	$Ca_3SiO_5 [C_3S]*$
C_3A	$Ca_3Al_2O_6 [C_3A]$
anhydrite	$CaSO_4$
Al-hauyne (sodalite structure)	$C_4A_3\bar{S}$
$12CaO \cdot 7Al_2O_3$	$C_{12}A_7$

*Cement nomenclature: $C = CaO$, $A = Al_2O_3$, $S = SiO_2$, $\bar{S} = SO_3$.

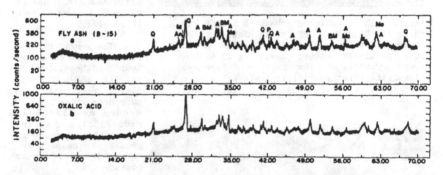

Fig. 3. X-ray Diffractogram of High-Calcium Fly Ash (B-15) Similar to that
in Fig. 1. (a) Before etching, (b) after 45 minutes oxalic acid, etch. (Q
= quartz, m = mullite, A = alite, BM = brownmillerite, Ma = magnetite, P =
periclase) [2].

Another common component of high calcium fly ashes, present frequently
in fairly substantial amounts, is C_3A. This compound can react rapidly with
any anhydrite present to generate ettringite, and self-setting high calcium
fly ashes are known where the setting is clearly due to rapid production of
ettringite rods that link and bind adjacent spheres.

C_2S and even C_3S are detected occasionally in high-calcium fly ashes,
but only in small amounts; their eventual hydration probably contributes
slightly to the formation of cementitious products.

Periclase (crystalline MgO) is reported in some fly ashes, in small
amounts. ASTM C 618 limits the analytical MgO content of fly ash to be used
in concrete to 5% or less, based on experience with portland cement where
$Mg(OH)_2$ formation may lead to unsoundness. However, high-calcium fly ashes
having up to 6% analytical MgO (present as periclase, merwinite, and in
glass) routinely show no unusual expansion in the ASTM autoclave expansion
test, so this requirement appears not to be appropriate for fly ash (O. Manz
and T. Demirel, private communication).

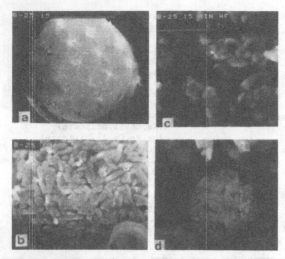

Fig. 4. SEM Photomicrographs of a Low-Ca Fly Ash (B-25) Etched in HF for 15 Minutes. (a) Sphere with a layer of noncrystalline solid (NCS) partially etched away, (b,c) iron oxide crystals on the surface of sphere, (d) mullite crystals [12]. Scale (a) 1 cm = 5μm, (b) 1 cm = 1μm, (c) 1 cm = 0.5μm, (d) 0.5 cm = 1μm.

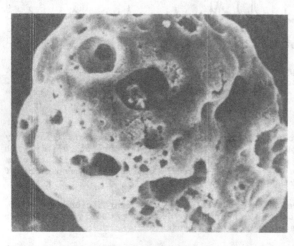

Fig. 5. Scanning Electron Micrograph of Vesicular Spherical Particle After HF Treatment, X370, Believed to be Largely or all Mullite [11].

Finally, gehlenite ($Ca_2Al_2SiO_7$) or melilite solid solutions (Ca, Mg, Al pyrosilicates) are sometimes observed in high calcium fly ashes: these are not ordinarily very reactive.

Because the crystalline phases of fly ashes are ordinarily embedded within the glassy spheres, their morphologies are not usually recognizable in the untreated fly ash. They are made visible by deliberate etching treatments to dissolve away some of the glass, or after the glass has reacted in concrete [2,5,7-11].

The x-ray diffraction pattern of a typical high-calcium fly ash is shown in Fig. 3(a). Here the broad amorphous glass maximum occurs at ~32°2θ (Cu radiation), and several crystalline phases are present as evident from the sharp diffraction peaks.

Selective dissolution techniques are often used to attempt to separate individual phases for analysis. Figure 3(b) shows the result of oxalic acid dissolution, which has removed much of the glass in a high-calcium fly ash [2]. Hydrofluoric acid (HF) dissolution of the glass in low-calcium fly ashes has been used quite successfully to separate a residue of crystalline components including mullite, quartz, and sometimes iron oxides from low-calcium fly ashes [2,7]. The SEM images of ashes resulting from one such treatment are shown in Fig. 4. A view of an individual residual mullite skeleton left after HF treatment of a low-calcium fly ash is shown in Fig. 5. Occasionally, fly ashes show spheres with substantial crystals of insoluble substances (as distinguished from soluble powders condensed from the vapor phase) on their surfaces, evident even without etching. Figure 6 shows such crystals, thought to be hematite, on the surface of spheres of an iron rich low-calcium fly ash. Lauf [7c] has further shown the detail which can be revealed through ion beam etching for TEM studies.

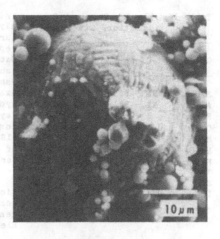

Fig. 6. Surface Deposition of Hexagonal (Hematite) Crystals on a Sphere in an Iron-Rich Low-Ca Fly Ash.

Particles of unburned coal residue occur in many fly ashes, and are common in those formed under poor burning conditions [13]. They occur in various morphologies, some of which are illustrated in Fig. 7 [11]. Unburned coal particles constitute a major fraction of the coarsest fraction of many fly ashes, although they are found in fine sizes as well. Irregular oversized particles composed mostly of vesicular glass are also common, and may not immediately be distinguishable from coarse unburned coal fragments. Many carbonaceous fragments, both oversize and fine, contain voids in which fine mineral fly ash spheres have been formed, apparently by melting and resolidification of the finely-divided clay and other mineral matter within the ground coal fragment.

It would be extremely desirable to have a procedure available for the quantitative determination of the amounts glass and of each of the crystalline components of a given fly ash. This is

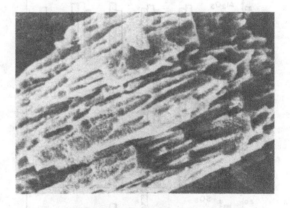

Fig. 7. Scanning Electron Micrograph of an Opaque Particle of Vesicular Carbon from High Lime Ash, X250 [11].

an extremely difficult undertaking, and a general procedure satisfactory for all fly ashes has not yet been developed. However, considerable success has been recently reported for such quantitative mineralogical analysis of certain low-calcium fly ashes [10].

Chemical Composition and Physical Properties

The variability of the chemical composition of fly ash is illustrated

Table IV. Fly Ash Chemical Composition, wt% Range.

Oxide	Class F	Class C
SiO_2	38 - 65	33 - 61
Al_2O_3	11 - 33	8 - 26
Fe_2O_3	3 - 31	4 - 10
CaO	0.6 - 13.3	14 - 37
MgO	0 - 5.0	1.0 - 7.0
Na_2O_3	0 - 3.1	0.4 - 6.4
K_2O	0.7 - 5.6	0.3 - 2.0
TiO_2	0.7 - 5.6	0.9 - 2.8
SO_3	0 - 4.0	0.5 - 7.3
LOI	0.1 - 12.0	0.2 - 1.4

in Table IV, which is derived from analyses reported throughout the literature. In almost all cases these are represented in United States fly ashes. Naturally, such variations in composition cause concern over performance of fly ashes in cement and concrete [15]. For example, reaction of the sulfates with hydrating aluminates can result in delayed setting and lower the resulting concrete strength. Thus, while ASTM C 618 limits the SO_3 content of both types of fly ash to 5%, some ashes have contents in excess of this limit (Table IV).

The variations in element distribution in sub-bituminous and bituminous coal ashes found within six different generating stations are shown in Fig. 8. Electrostatic precipitator ashes from air and boiler aside, and bottom ash are contrasted from the same power plant and coal source. There were no consistent trends, except for iron-rich ashes which showed a marked concentration of Fe in the bottom ash.

Unburned carbon particles are relatively inert in cement hydration reactions [1], acting mostly as a diluent to the system. However, if present in substantial quantity, they can adversely affect the workability, and lead to lower concrete strength [23]. Usually the high-calcium fly ashes have low carbon contents, while the carbon contents of low-calcium fly ash are somewhat more variable and higher. Carbon content is reflected in the loss on ignition (LOI) values, with typical ranges given in Table IV.

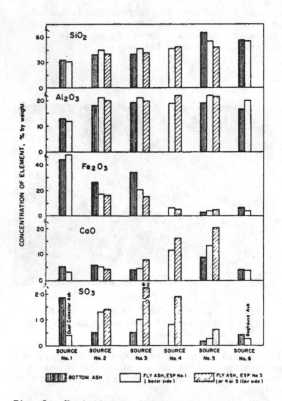

Fig. 8. Variation in SiO_2, Al_2O_3, Fe_2O_3, CaO and SO_3 in Different Thermal Regimes [23].

Specific gravity and fineness are usual physical properties reported. The specific gravity of fly ash varies considerably ranging from 1.6 to 2.8, related partly to its chemical composition and morphology. Fly ashes with a high iron content tend to have high specific gravities while those high in carbon have low specific gravities, about 1.6 to 2.0. High-calcium fly ashes have few cenospheres, generally have specific gravities of 2.4 to 2.8, and usually have finer particle sizes than low-calcium fly ashes. Typical ranges of particle size of high- and low-calcium fly ashes measured by hydrometer (ASTM D 422) are given in Fig. 9. Partitioning was pronounced in some cases: two electrostatic precipitators collected 80 to 90% of their particulates in the size range below 10 micrometers.

Fly Ash Classification

Because of the wide range of chemical compositions and associated behavioral characteristics of fly ashes, it is necessary to have some means of classifying them into subcategories.

An older classification of fly ashes based on the morphology of the fly ash particles as it relates to both original coal mineral characteristics and to combustion should be mentioned [25,26]. This classification, given in Fig. 10, was developed primarily for environmental studies rather than concrete utilization.

The 'natural' distinction between the broadest groupings would seem to be on the basis of their CaO content. High-calcium fly ashes are now known to consist primarily of glass of different structure and apparently higher reactivity than low-calcium fly ashes, and their suite of crystalline components includes calcium-bearing compounds that may be active in cementing reaction. In contrast, low-calcium fly ashes typically contain siliceous, relatively slowly reacting glass that hydrates in the presence of alkalis or lime, and crystalline components that remain inert in the concrete environment. This difference is recognizable in concrete behavior, and has given rise to statements in the literature that high calcium fly ashes have both pozzolanic and cementitious properties, while low calcium rly ashes are merely pozzolanic.

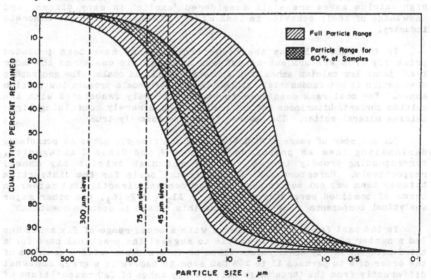

Fig. 9. Particle Size Analysis of High-Ca and Low-Ca Fly Ashes [23].

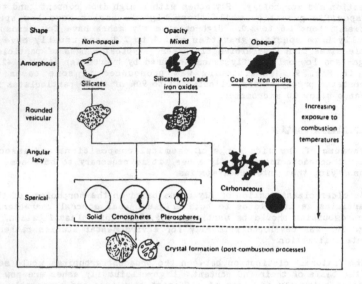

Fig. 10. Fly Ash Morphogenesis Scheme Illustrating Probable Relationship of Opacity to Particle Combustion and Relationship of Particle Shape to Exposure to Combustion Temperatures (modified from [25,26]).

Low-calcium fly ashes have been produced and utilized for many years in the United States, and in many countries in Europe and elsewhere are the only type of fly ash being produced and used in practice. The more active high-calcium ashes are still considered 'exotic' in many places, and knowledge of their behavior is less widely disseminated in the concrete industry.

In the United States the high-calcium ashes have been produced primarily from Western sub-bituminous and lignitic coals and the more traditional low-calcium ashes from Eastern bituminous coals. The geographic distinction is not fundamental, since many Western coals produce low calcium ashes. The coal rank does seem to be more closely associated with the calcium content, bituminous and anthracitic coals rarely containing high-calcium mineral matter. The reverse is not necessarily true.

For a number of years ASTM C 618 has classified fly ashes as pozzolans, designating them as pozzolans of Class F and Class C categories, corresponding broadly to low calcium and high calcium fly ashes, respectively. Unfortunately, the numerical basis for the distinction between them was not set in terms of CaO content directly, but rather in terms of combined percentages of SiO_2 + Al_2O_3 + Fe_2O_3, the other major analytical components. The logic behind this scheme is open to question.

In the past few years experience with a broad range of fly ashes has led a number of knowledgeable people to suggest the practical need for a third category of fly ashes, since ashes of intermediate calcium contents of the order of 8 to perhaps 15 to 20% CaO seem to behave as a group somewhat differently from the 'true' high calcium fly ashes of CaO compositions of the order of 25% CaO and more.

Suggestions toward a revised classification scheme for concrete use have been published by several authors. Diamond [ref. 3, p. 6-15] proposed several chemical and physical criteria which might be involved in an adequate classification.

One particular scheme [24] would reserve high-calcium fly ashes within a certain compositional range for uses when fly ash replacement in concrete exceeds 25% to insure development of adequate strength in the concrete, leaving fly ashes out of this range for use in less-critical lower replacement percentage uses.

Finally, on a somewhat different level, a proposal for classifying United Kingdom fly ashes, all of which are low-calcium ashes, into categories based on the mineral chemistry of their coal sources has recently been published [27]. The ratio of K to Al in the coal seems to provide a good indication of the fluxing action of K_2O in fly ash production, hence a good correlation to glass content and presumably to comparative reactivity in concrete (Fig. 11). The higher K/A ratios generally have higher reactivity. However, composition is certainly not the only factor, as granulometry is also very important.

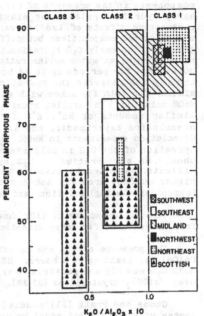

Fig. 11. Regional Classification of U.K. Fly Ashes on the Basis of Their PPI [Pozzolanic Potential Index, = K_2O/Al_2O_3 (wt) x 10].

FLY ASH REACTIONS WITH CEMENT AND CONCRETE COMPONENTS

Reactions of Low-Calcium Fly Ashes

In general terms, the chemical reactions of low-calcium fly ashes in concrete involve hydration of the siliceous glass (including its Al_2O_3 and probably its Fe_2O_3 constituents as well) and reaction with the components released from the hydrating cement system, including but not limited to calcium hydroxide and alkalis. Roy [ref. 3, p. 5.1-5.5] discussed the energetics of hydration of metastable glasses such as those in fly ashes when they react with $Ca(OH)_2$ or alkalis to produce cementitious products. The net result is a reduction in the amount of calcium hydroxide in the products, and presumably an increase in the amounts of C-S-H gel and of calcium aluminate hydrates produced; the increases are difficult to document.

Incorporation of the fly ash may induce changes in the cement hydration rates; in any event the early strengths of the concretes are ordinarily reduced, and strengths at very late ages usually increased. The age at which the strength curves cross each other varies considerably with the system being considered, and has been reported to be anywhere from 28 days to as much as six months, depending partly on curing conditions [3,28].

Recent studies on the hydration of cement and of individual cement components in the presence of fly ash indicate that fly ash may influence early hydration rates. For example, Jawed and Skalny [29] studied the comparative effects of two low-calcium fly ashes of similar chemical and mineralogical compositions but different surface areas and apparent specific gravities, on early C_3S hydration. C_3S and C_3S + fly ash (20% replacement) were hydrated at water:solids ratios of 0.50 using either water or 0.5N NaOH solutions, for periods up to 24 hours. It was shown that the presence of the fly ash retarded the hydration of C_3S; the retardation effect being similar for both fly ashes with water, but different on hydration with the NaOH solution. In parallel studies of the two fly ashes, it was found that similar amounts of Na^+, K^+, Ca^{2+}, and SO_4^{2-} were dissolved in water dissolution experiments; and in both cases Si and Al were detected in dissolution experiments in NaOH. Subsequently, studies were carried out of hydration of the C_3S in silicate and aluminate-bearing solutions. These showed an acceleration of C_3S hydration in the presence of dissolved silicate, but a deceleration in the presence of dissolved aluminate. The higher surface area fly ash dissolved more Al in the alkali-containing (cement-simulating) solution, and caused greater retardation.

Tenoutasse and Marion [30] showed that essentially all the sulfate from four low-calcium fly ashes dissolved in water in 60 minutes.

The presence of fly ash is often observed to delay crystallization of $Ca(OH)_2$ at least up to 3 hours. SEM observations at 6–8 hours showed that both C_3S and fly ash particles may be covered with C–S–H fibers deposited over $Ca(OH)_2$ crystallites [31,32].

Ghose and Pratt [31] studied hydration reactions in cement-fly ash pastes with 30% fly ash added by weight, and water:solids ratios of 0.38 and 0.50. Calorimetric experiments indicated that fly ash increased the induction period (from about 2 hours to almost 3 hours for w:s = 0.5 pastes). It also increased the time to reach the peak of the main hydration exotherm and attenuated the shoulder on this peak usually associated with ettringite-monosulfate conversion. An additional heat evolution peak, possibly associated with pozzolanic activity, was recorded at about 1–2 days. Fly ash appeared to act to retard both silicate and aluminate hydration. Detailed SEM observations paralleled the calorimetric studies. It was concluded that the presence of the fly ash influences cement hydration but that the reaction of the fly ash itself was slow and variable between fly ash particles, with some reacting within 1 day and others not even after 5 months.

Gutteridge and co-workers [10] used quantitative XRD methods mentioned previously to investigate cement-fly ash hydration. The glass contents of a suite of fly ashes were first determined by differences of the sum of the crystalline components from 100%. The fly ash with the highest glass content was blended with portland cement at a 30% replacement level and hydration of this blended cement was compared with hydration of the portland cement alone. In this system the rate of alite hydration was found to be increased by the fly ash. Fly ash was determined by selective dissolution techniques as residual from the hydrating pastes and examined by the quantitative XRD procedures; as expected it was found that the residual content of fly ash progressively diminished as hydration proceeded. It was found that a weight of 'glass' equivalent to 42% of the original weight of the fly ash had dissolved in one year.

Ghose and Pratt [31] and Halse et al. [9] showed that relatively little retardation was actually produced in some cement-fly ash combinations, as indicated in the calorimetric curves of Fig. 12. Similar observations are reported elsewhere in this symposium [32].

Rayment [33] investigated the reaction products produced in a high alkali (K_2O) content cement paste with and without a 20% replacement of a low-calcium fly ash. Room-temperature curing of the w:s 0.50 pastes was carried out for 8 days, after which they were immersed in isopropyl alcohol, vacuum dried, and polished sections prepared for study with an x-ray microanalyzer. It was indicated that the fly ash addition lowered the C:S ratio of the C-S-H gel produced, and increased its potassium content. The lowered C:S ratio was construed as an increase in silicon content rather than a decrease in calcium content of the gel product.

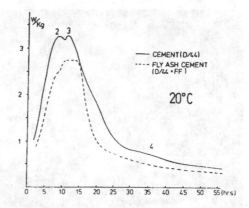

Fig. 12. Rate of Heat Output for Cement D44, Containing 69% Alite and its Mixture with 20% of a Low-Ca Fly Ash (Fiddler's Ferry).

Techniques for determining the degree of reaction of fly ash at various times in fly ash-cement systems are important. Ohsawa et al. [34] studied the hydration of model systems composed of a low calcium fly ash mixed with gypsum and lime, and determined the residual fly ash content after dissolution with a picric acid-methanol solution. They concluded that such a procedure, satisfactory with the model system, would also be satisfactory with cement-fly ash systems.

Finally, Mohan and Taylor [35] investigated the effects of a high glass content low-calcium fly ash on the reaction of C_3S pastes at a 25% replacement level. The pastes were prepared at w:s = 0.50 and cured for up to 397 days at room temperature. After drying at room temperature under vacuum the residual content of C_3S was determined by XRD and the amount of unreacted fly ash by salicylic acid-methanol extraction. Elemental ratios in the products were determined by analytical electron microscopy and contents of $Ca(OH)_2$ and bound CO_2 by thermogravimetry. It was found that fly ash accelerated the C_3S hydration, but progressively reduced the content of $Ca(OH)_2$ and eliminated it after about 90 days. The mean Ca:Si ratio of the C-S-H gel particles examined individually was reduced from 1.51 to 1.43, but this reduction was independent of time. The fly ash itself was slowly but progressively dissolved, about 15% of it having reacted at 28 days and 45% at 1 year.

Comparative Effects of High-Calcium and Low Calcium Fly Ashes

Most of the prior section described results obtained with low-calcium fly ashes. The influences of high-calcium fly ashes may be expected to be somewhat different, and perhaps more intense in view of their greater reactivity.

It has been found [32] that the early retardation of cement hydration [including $Ca(OH)_2$ and C-S-H nucleation] was more severe with a high-calcium than with a low-calcium fly ash. The total heat evolved from equivalent masses is of course greater in high-calcium fly ashes mixtures in the first

16

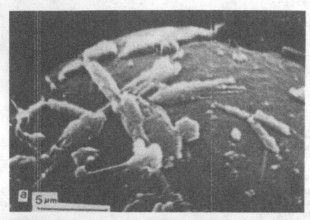

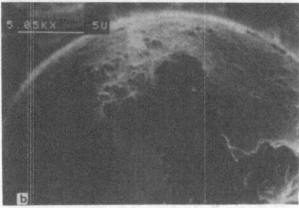

few days. Studies involving calorimetry, solution analysis, and SEM analysis have indicated the complexity of the reactions taking place. In one study [36] definite evidence of the dissolution of some of the glass phase was found after only 3 days. Cenospheres in particular were found to react quickly, the glass reacting to leave a skeleton of crystalline material as indicated in Fig. 13(b).

Comparative effects of different low-calcium and high-calcium fly ashes in the ASTM pozzolanic index test with cement were studied by Berry and Hemmings [23]. As indicated in Fig. 14, members of both groups passed the tests, and other members failed the test.

Mehta [19] was concerned with test results showing low strengths of low-calcium fly ash mixtures. He showed that when accelerated curing (at 43°C) was carried out for 7

Fig. 13. (a) High Lime Fly Ash After Hydration 3 Days, 38°C in Complex Cementitious Mixture [40] (bars = 1μm). (b) Same Ash as in (a), Showing Essentially Complete Reaction of Cenosphere After 3 Days.

days, the strengths for low calcium fly ashes were nearly as high as for high-calcium fly ashes, and that accelerated test results at 7 days were equivalent to normal curing at 28 days (Fig. 15). Similar comparisons were made between 28-day accelerated curing and 90-day room temperature curing results.

Finally, Coule [37] used a calorimeter to simulate temperature rise in concrete pours of different size; he found that expected temperature rise in a 3 m³ pour would reach a 55°C maximum at about 72 hours, justifying the opinions expressed by Mehta [19] and others [22,28] that large masses cause significant heat build-up and consequently earlier strength development. Similar results have been found in other complex blended cements containing fly ash [38].

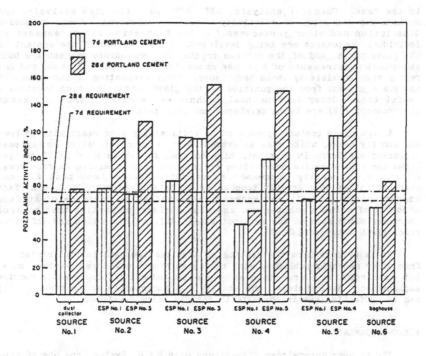

Fig. 14. Variation in Pozzolanic Index for Fly Ashes. Numbers 4 and 5 are high-CaO, number 3 contains 7% CaO, others are unusual low calcium [29].

CONCLUSIONS

This paper has been aimed at providing an analysis of the results of recent investigations on fly ash characterization and its relevance to performance of fly ash in concrete. Much of the work reported in the literature has not yet attempted to assess effects in concrete per se, but is restricted to cement paste and mortars, and even individual cement component systems.

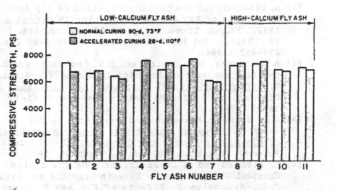

Fig. 15. Compressive Strength of Mortars Containing Portland Cement-Fly Ash Mixtures [19].

Nevertheless, despite the obvious complexity of fly ashes as materials, methods of characterization are being developed that appear to be adequate to the task. Chemical analysis, XRD, SEM, particle size analysis, and methods for the study of reactivity have been discussed. Newer selective dissolution and other procedures for the separation and assessment of individual components are being developed. Determination of the content of the glass phase, and of the various crystalline components present are both important requirements of a proper characterization of a fly ash, and are partly made possible by these techniques. X-ray estimation of the nature of the glass present from the position of the glass scattering hump provides a useful tool. Other instrumental techniques, such as microfocus Raman spectroscopy [38] are being developed and show promise.

A high glass content appears to be critical for good reactivity in low-calcium fly ashes, which show an extensive range of reactivities as has been discussed earlier. In contrast, high-calcium fly ashes have both reactive glass and some reactive crystalline components. High specific surface areas of course, generally increases reactivity. Other important fly ash characteristics only briefly touched on in this discussion include surface characteristics of fly ash spheres, especially the presence of surface deposits of soluble substances, and also surface properties that control characteristics such as the zeta potential [41], which in turn may influence rheological behavior [40,42].

It is evident from the results of previous Symposia in this series and from other recently-published work that significant advances have been made, but much remains obscure and many unresolved conflicting results require explanation. Nevertheless, the ability to characterize and categorize fly ash for effective use in concrete is rapidly being developed.

ACKNOWLEDGEMENT

The authors acknowledge discussions with H.F.W. Taylor, and one of them (DMR) acknowledges partial support by NSF grant CPE-8112821.

REFERENCES

[1] S. Diamond, (a) The Characterization of Fly Ashes, in Proc. Symposium N, Effects of Fly Ash Incorporation in Cement and Concrete, Boston, 1981. Ed., S. Diamond, pp. 12-23; Mater. Res. Soc., University Park, PA (1982),. (b) The Utilization of Fly Ash, Cem. Concr. Res. 14 (4), 455-462 (1984).

[2] B.E. Scheetz, D.W. Strickler, M.W. Grutzeck, and D.M. Roy, Physical and Chemical Behavior of Selectively Etched Fly Ashes, in Proc. Symposium N, Effects of Fly Ash Incorporation in Cement and Concrete, Ed. S. Diamond, pp. 24-33, Materials Research Society, University Park, PA 16802 (1982).

[3] G.M. Idorn. Research and Development for the Use of Fly Ash in Concrete, Workshop Proceedings: Research and Development Needs for Use of Fly Ash in Cement and Concrete, EPRI CS-2616-SR, Electric Power Research Institute, Palo Alto, CA (Sept. 1982).

[4] S. Diamond and F. Lopez-Flores. On the Distinction in Physical and Chemical Characteristics Between Lignitic and Bituminous Fly Ashes, in Proc. Symposium N, Effects of Fly Ash Incorporation in Cement and Concrete, Boston, 1981. Ed., S. Diamond, pp. 34-44. Materials Research Society, University Park, PA (1982).

[5] S. Diamond. On the Glass Present in Low-Calcium and in High-Calcium Fly Ashes. Cem. Concr. Res. 13 (4), 459-464 (1983).

[6] S. Diamond. Characterization and Classification of Fly Ashes in Terms of Certain Specific Chemical and Physical Parameters, 9-20 in The Use of PFA in Concrete, Ed. J. Cabrera and A. Cusens, Dept. of Civil

Engineering, Leeds, Concrete Society and Central Electricity Generating Board, U.K. (1982).

[7] (a) L.D. Hulett, A.J. Weinberger, N.M. Ferguson, K.J. Northcutt and W.S. Lyon. Trace Elements and Phase Relations in Fly Ash. EA-1822. Res. Proj. 1062, Final Report for Electric Power Research Institute, Palo Alto, CA (May 1981). (b) L.D. Hulett, Jr. et al., Science 210 (19), 1356 (Dec. 1980). (c) R. Lauf, Microstructures of Coal Fly Ash Particles, Ceramic Bulletin 61, 487-490 (1982).

[8] G.J. McCarthy, K.D. Swanson, L.P. Keller, and W.C. Blatter, Mineralogy of Western Fly Ash, Cem. Concr. Res. 14, 471-478 (1984).

[9] Y. Halse, P.L. Pratt, J.A. Dalziel and W.A. Gutteridge, Development of Microstructure and Other Properties in Fly Ash OPC Systems. Cem. Concr. Res. 14, 491-498 (1984).

[10] W.A. Gutteridge. Quantitative X-ray Powder Diffraction in the Study of Some Cementive Minerals, in The Chemistry and Chemically-Related Properties of Cement, Ed. F.P. Glasser, British Ceramic Society, Shelton, Stoke-on-Trent (1984).

[11] A.D. Buck, T.B. Husbands and J.P. Burke, Studies of the Constitution of Fly Ash Using Selective Dissolution, Misc. Paper SL-83-5, U.S. Army Engineer Waterways Experiment Station (May 1983).

[12] P.L. Pratt. The Influence of Pulverized Fuel Ash on the Hydration of Cement and Concrete, 29-51 in The Use of PFA in Concrete, Ed. J. Cabrera and A. Cusens, Dept. of Civil Engineering, Leeds, Concrete Society and Central Electricity Generating Board, U.K. (1982).

[13] S. Diamond. Intimate Association of Coal Particles and Inorganic Spheres in Fly Ash, Cem. Concr. Res. 12, 405-407 (1982).

[14] S. Schlorholtz and T. Demirel, Determination of Free Lime (CaO) in Fly Ashes (this Symposium).

[15] Standard Specifications for Fly Ash and Raw or Calcined Natural Pozzolan for Use as a Mineral Admixture in Portland Cement Concrete (ASTM C 618-80 and ASTM C 311-77). 1982 Annual Book of ASTM Standards, Part 14: 1984, Vol. 04.02.

[16] E.L. White, M. Lenkei, D.M. Roy, and F.D. Tamas. Fly Ash Slurries with Superplasticizers (this Volume).

[17] D.J. Thorne and J.D. Watt. Investigation of the Composition, Pozzolanic Properties and Formation of Pulverized Fuel Ash, British Coal Utilization Research Association, Information Circular No. 265.

[18] P.K. Mehta and O.E. Gjørv. Properties of Portland Cement Concrete Containing Fly Ash and Condensed Silica Fume, Cem. Concr. Res. 12 (5), 587-595 (1982).

[19] P.K. Mehta. Testing and Correlation of Fly Ash Properties with Respect to Pozzolanic Behavior. EPRI1, CS-3314, prepared for Electric Power Research Institute, Palo Alto, CA (1984).

[20] S. Diamond. Effects of Two Danish Fly Ashes on the Alkali Contents of Pore Solutions of Cement-Fly Ash Pastes, Cem. Concr. Res. 11 (3), 383-394 (1981).

[21] B.B. Hope. Autoclaved Concrete Containing Fly Ash, Cem. Concr. Res. 11 (2), 227-233 (1981).

[22] Proc. Symposium N, Effects of Fly Ash Incorporation in Cement and Concrete, Boston, 1981. Ed., S. Diamond, pp. 26, 47, 61, 73, 82, 93, 103,135, 186, 281, 292, 297, Materials Research Society, University Park, PA (1982).

[23] E.E. Berry and R.T. Hemmings. Coal Ash in Canada: Vol. 2, Laboratory Evaluation of Coal Ash, Report for Canadian Electric Association (March 1983).

[24] O.E. Manz. American and Foreign Characterization of Fly Ash for Use in Concrete. Proc. Symposium N, Effects of Fly Ash Incorporation in Cement and Concrete, Boston, 1981. Ed., S. Diamond, pp. 269-279, Materials Research Society, University Park, PA (1982).

[25] W.R. Roy, R.G. Thiery, R.M. Schuller, and J.J. Suloway. Coal Ash: A Review of the Literature and Proposed Classification System with

Emphasis on Environmental Impacts. Environmental Geology Notes 96, Illinois State Geological Survey, Champaign, IL, Apr. 1981.

[26] G.L. Fisher, et al. Physical and Morphological Studies of Size Classified Coal Fly Ash: Environmental Science and Technology 12 (4), 447-451 (1978).

[27] F.H. Hubbard, R.K. Dhir, and M.S. Ellis. Pulverized-Fuel Ash for Concrete: Compositional Characterization of United Kingdom PFA. Cem. Concr. Res. 15, 185-198 (1985).

[28] (a) G.M. Idorn and K.R. Henriksen. State of the Art for Fly Ash Uses in Concrete. Cem. Concr. Res. 14, 463-470 (1984). (b) Ref. [22], pp. 244-259. (c) Thirty Years of Alkalis in Concrete, in Alkalis in Concrete, Proc., Danish Concrete Assn., Copenhagen, pp. 21-38 (1983).

[29] I. Jawed and J. Skalny. Hydration of Tricalcium Silicate in the Presence of Fly Ash. Proc. Symposium N, Effects of Fly Ash Incorporation in Cement and Concrete, Boston, 1981. Ed. S. Diamond, pp. 60-69, Materials Research Society, University Park, PA (1982).

[30] N. Tenoutasse and A.M. Marion. Influence of Fly Ash in the Structure of OPC and Pure Calcium Silicates, pp. 359-374 in The Chemistry and Chemically-Related Properties of Cement, Ed. F.P. Glasser, British Ceramic Society, Shelton, Stoke-on-Trent (1984).

[31] A. Ghose and P.L. Pratt. Studies of the Hydration Reactions and Microstructure of Cement-Fly Ash Pastes. Proc. Symposium N, Effects of Fly Ash Incorporation in Cement and Concrete, Boston, 1981. Ed., S. Diamond, pp. 82-91, Materials Research Society, University Park, PA (1982).

[32] M.W. Grutzeck, Wei Fajun, and D.M. Roy. Retardation Effects in the Hydration of Cement-Fly Ash Pastes (this Volume).

[33] P.L. Rayment. The Effect of Pulverized-Fuel Ash on the C/S Molar Ratio and Alkali Content of Calcium Silicate Hydrates in Cement. Cem. Concr. Res. 12 (1), 133-140 (1982).

[34] S. Ohsawa, K. Asaga, S. Goto, and M. Daimon. Quantitative Determination of Fly Ash in the Hydrated Fly Ash-$CaSO_4 \cdot 2H_2O-Ca(OH)_2$ System, Cem. Concr. Res. 15, 357-366 (1985).

[35] K. Mohan and H.F.W. Taylor. Pastes of Tricalcium Silicate with Fly Ash--Analytical Electron Microscopy, Trimethylsilylation and Other Studies. Proc. Symposium N, Effects of Fly Ash Incorporation in Cement and Concrete, Boston, 1981. Ed., S. Diamond, pp. 54-59, Materials Research Society, University Park, PA (1982).

[36] M.W. Grutzeck, D.M. Roy, and B.E. Scheetz. Microstructures of High-Lime Fly Ash Cementitious Mixes. Cem.Concr. Res. 11 (2), 291-294 (1981).

[37] M.J. Coule. Calorimetric Studies of the Hydration Behavior of Extended Cements, pp. 385-401 in in The Chemistry and Chemically-Related Properties of Cement, Ed. F.P. Glasser, British Ceramic Society, Shelton, Stoke-on-Trent (1984).

[38] C. Gotsis and D.M. Roy. Parametric Analysis of the Heat Evolution During Hydration of a Cementitious Plug in a Borehole, Cem. Concr. Res. 14, 847-854 (1984).

[39] W.B. White and B.E. Scheetz. Characterization of Crystalline Phases in Fly Ash by Microfocus Laser Raman Spectroscopy (this Volume).

[40] D.M. Roy, M.W. Grutzeck, and L.D. Wakeley, Selection and Durability of Seal Materials for a Bedded Salt Repository: Preliminary Studies, ONWI-479, prepared by The Pennsylvania State University, for Office of Nuclear Waste Isolation, Battelle Memorial Institute, Columbus, OH (1983).

[41] R.I.A. Malek and D.M. Roy, Electrokinetic Phenomena and Surface Characteristics of Fly Ash Particles (this Volume).

[42] D.M. Roy, S. Diamond, and J.P. Skalny. The Effects of Mineral Admixtures on Rheological Characteristics of Cement Paste, in Proc. Sympoisum M, Concrete Rheology. Ed. Jan Skalny, pp. 152-173, Materials Research Society, University Park, PA (1982).

TRANSMITTED AND REFLECTED VISIBLE LIGHT
MICROSCOPY OF TWO BITUMINOUS FLY ASHES

DONALD L. BIGGS and JOSEPH J. BRUNS*
Department of Earth Sciences and Ames Laboratory, Iowa State University,
Ames, IA 50011

(Received 13 November, 1984; accepted 22 February, 1985; Refereed)

ABSTRACT

Fly ashes of high magnetic content taken from two midwestern power
plants were examined to determine the mineralogy of the magnetic and nonmagnetic fractions. Fly ash spheres from the magnetic fraction are predominantly composed of ferrite spinel, hematite and silicate glass. The
hematite appears to be a replacement product of the original ferrite spinel.
Nonmagnetic phases include mullite, lime, small amounts of hematite and
silicate glass. Quartz morphology indicates that it did not fuse in the
furnace. Mullite and lime have morphologies indicative of crystallization
in the furnace. Hematite is bonded to the nonmagnetic particles or as a
complete replacement of ferrite spinel spheres.

INTRODUCTION

The constantly increasing usage of coal in the United States is
accompanied by a steady increase in fly ash production. This material, the
result of coal combustion, is composed of the mineral matter originally
present in the coal more or less altered by the environment of the furnace
and mixed with unburned, or partially burned coal particles. Some of the
minerals present in the coal as it is fed into the furnace are unaltered
and proceed through the furnace as finely divided particulate matter. The
chief constituent of which this is true is quartz. Most crystalline species
present in the coal as fed into the furnace undergo reactions of varying
sorts and the end result is the production of a finely divided material
composed principally of oxides of iron, aluminum, titanium, other elements
and silicate glass more or less intimately mixed with each other and small
amounts of unburned coal. The resolution characteristics of visible region
microscopy are well suited to the study of chemical and spatial relationships encountered in the several phases found in fly ash. Moreover,
optical properties of most crystalline phases in this region are well
understood.

The usage of fly ash has been well summarized by D. M. Roy, K. Luke,
and S. Diamond in this volume. This study is directed toward understanding
the mineralogy of fly ashes and the evaluation of the material as a source
of economically important elements such as iron, aluminum and titanium.

Fly ash specimens for use in this study were obtained by the Ames
Laboratory and the Departments of Chemical and Nuclear Engineering of

*Present address, Beard Oil Company, 503 Park Harvey Center, Oklahoma City,
Oklahoma 73102
**Operated for the U.S. Department of Energy by Iowa State University under
contract No. W-7405-Eng-82. This work was supported by the Assistant
Secretary for Fossil Energy, Office of Coal Mining, through the Pittsburgh
Mining Technology Center, Coal Preparation Branch.

TABLE I
Fly ash reference names and sources of specimens

Reference Name	Plant and Location
KA	Louisville, Plant, Louisville, Kentucky
MA	Montrose Plant, Kansas City, Missouri

Iowa State University. Two fly ashes from midwestern bituminous coals were selected for the experiments. Both had been subjected to previous investigations concerning extraction of some or all the contained elements (1, 2, 3, 4, 5, 6, 7, 8). Table I gives the name and location of the specimens.

The fly ashes were separated into a magnetically enriched and a magnetically impoverished fraction in a Roto-Flux Magnetic Separator (9). This instrument is designed to produce a rotating magnetic field that develops a spinning motion in and indirectly effects a transporting movement of magnetically attracted particles.

The mineralogy of the fractions was determined by powder method X-ray diffraction techniques supplemented by wet quantitative analysis. The X-ray experiments were performed on a Picker theta-theta diffractometer. Because the patterns derived from these experiments were patterns of mixed powders, some overlap of diffraction maxima and alteration of intensity values was unavoidable. Therefore, the acceptance of a diffraction maximum as unambiguously representative of a compound required that two other maxima characteristic of the species occur in the same pattern and in the correct intensity. Wet quantitative analysis was performed as a check on the X-ray determinations and to learn the effect of composition on size of ash particle. Previous workers (2, 7, 8) have reported the chemical analyses for the raw ashes before sizing.

Specimens were examined in the visible region of the electromagnetic spectrum using a Zeiss Photomicroscope Pol I equipped for transmitted and reflected light studies. Specimen preparation for these studies consisted of mounting small aliquots of fly ash in resin cylinders. After the resin had cured the cylinders were ground and polished on the end containing the most fly ash. Grinding and polishing were performed on standard grinding and polishing laps. Final polish was through Linde A and B grades of levigated alumina. The polished blocks were examined in polarized reflected light using oil immersion techniques. Nonopaque and nonmagnetic crystals were studied in thin section and loose crystal mounts immersed in oil.

DISCUSSION OF RESULTS

Magnetic Separation

The Roto-Flux Magnetic Separator is designed for larger loads than we were able to use and the separation was not as efficient as we desired. Using the loadings we were able to provide the instrument, we found it desirable to reduce the setting of the magnet control to 60% of full power. At this setting, we found that the magnetically enriched fraction contained less nonmagnetic material and the magnetically impoverished sample contained less material attracted to the magnet than when the instrument was operated at full power. We attribute this to the simple action of entrapment while the sample was going through the field at such a rate as to lower the probability of clean separation. Nevertheless, the instrument functioned well enough for the purpose of concentrating magnetically attractive and nonmagnetically attractive phases so that they could be studied.

X-ray Diffraction Analysis

Crystalline components identified in this way were: quartz, mullite, lime, hematite and ferrite spinel. Mullite, hematite, and ferrite spinel patterns were characterized by broad diffraction maxima. These characteristics are attributed to the high viscosity of the aluminum-rich silicate from which mullite is considered to have crystallized and to solid solution occurring in the ferrite spinel phase, or perhaps to the occurrence of one or more of the ferrite spinel compositions having slightly different unit cell sizes in intimate association with magnetite. The latter possibility seems very likely inasmuch as maghemite is also magnetic, unstable tending to invert to hematite and has a spacing a_0 of 8.34 as compared to that of magnetite, 8.396 Å. If maghemite indeed occurs in intimate association with magnetite, then the inversion to hematite would explain some of the observations in visible region microscopy.

Chemical Analysis

Chemical analyses were performed to show the effect of particle size on mineralogy in fly ash formation. Though that part of the investigation is not reported here, the information derived from these analyses was useful to check the determinations from the X-ray diffraction experiments. No elements were encountered in the analyses that were not represented in the phases identified by diffraction techniques or probable components of the glass phase.

Microscopy of the Magnetically Enriched Fraction

X-ray diffraction and chemical analyses indicated that the magnetic fractions of KA and MA fly ashes were intimately intergrown or fused mixtures of ferrite spinel, hematite, glass, and quartz. The microscopic investigation was undertaken to more precisely describe the internal structure and morphology of magnetic ash particles, which when viewed in reflected light "exhibit separate phases, grain boundaries, twinned crystals and voids" (2).

Initial observations showed that most particles were individual spheres consisting of three phases: ferrite spinel, hematite, and glass in varying proportions. Figure 1 is a representative field of fly ash particles. Note the predominantly spherical character and the diversity of species. Some magnetic particles are composite particles consisting of two or more components (Fig. 2, 3).

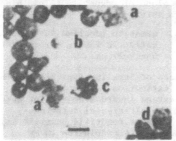

Figure 1. A typical assemblage of fly ash particles a and a' composite particles; b fusion of magnetic and nonmagnetic particles; c irregular shaped particles; d fragment of magnetic cenosphere, transmitted light oil immersion; bar=120 μm

Ferrite spinel and hematite were differentiated on the basis of optical properties and crystal habit. Ferrite spinel is optically isotropic with low to medium reflectance and appears grayish to blue gray in vertically incident light. Hematite is concentrated in the magnetic fraction because of its close association with ferrite spinel rather than its own magnetic attraction. Hematite has medium reflectance and appears

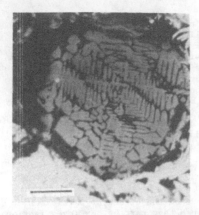
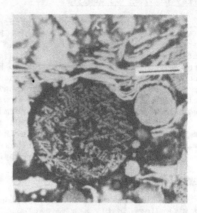

Figure 2. Dendritic growth of ferrite Figure 3. Network of small ferrite
           spinel in glass. Reflected   spinel dendrites in glass.
           light, oil immersion;   Reflected light, oil im-
           bar = 25 μm.   mersion; bar = 25 μm.

light bluish-gray in vertically incident light. The crystal is anisotropic,
bireflectant, and, under cross polarizers, brownish-gray to nearly extinct
in the 0° position and light bluish-gray in the 45° position.

    Two crystal habits are common in ferrite spinel, octahedra and den-
dritic masses. Dendrites are by far the most abundant form and they are
intergrown with a dark isotropic matrix considered to be silicate glass.
The coarseness, length, and complexity of the branches in the dendritic
forms are highly variable. Lengths and widths of branches vary from less
than one to tens of μm. Intricate networks of delicate branches are common
(Fig. 3) as are those having coarse and less numerous branches (Fig. 2).
The dendritic habit is considered indicative of rapid crystallization, an
interpretation in good agreement with the dark, transparent, isotropic
matrix which is interpreted as silicate glass.

    The most distinct octahedral cross-section observed was an equilateral
triangular face, considered to be the (111) plane of magnetite. This plane
is also thought to host lamellae of hematite intersecting at 60° (Fig. 4).

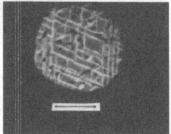

Figure 4. Hematite lamellae in spinel.
           Reflected light, oil immersion;
           bar = 25 μm.

Parallel sets of lamellae typically exhibit
concurrent extinction. Most lamellae
appear to extend inward from a point originating at the air-crystal interface
(Fig. 5). Rims of dendritic hematite and
coarse lamellae are considered to be
pseudomorphs of ferrite spinel. Pseudo-
morphic hematite, sometimes called martite,
and the occurrence of lamellae which extend inward from the air-crystal
interface along the (111) planes of ferrite spinel indicate that hematite is
an oxidation product of ferrite spinel or inversion of maghemite to hematite.
The inversion hypothesis appears less likely than that of oxidation because
of the coincidence of hematite lamellae and surfaces that are thought to be

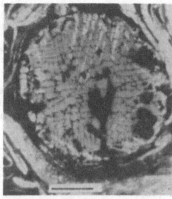

Figure 5. Hematite lamellae growing inward from ferrite spinel
surface. Reflected light,
oil immersion; bar = 25 μm

crystallographically controlled by the
ferrite spinel structure.

Oxidation of ferrite spinel is
thought to occur as free oxygen at the
surface of a crystal is ionized by
electrons derived from oxidation of
ferrous ions within the crystal. The
unbalanced charge within the original
crystal is restored by removal of 2 Fe^{3+}
ions for every 3 oxygens added at the
surface. It is thought that the Fe^{3+}
ions may diffuse through the close-packed oxygen framework to the surface
of the crystal where the Fe^{3+} ion and newly formed O^{2-} ion may be epitaxially
deposited as hematite on the crystal surface (5,6). At temperatures less
than about 600°C, a solid solution series between a cation deficient spinel,
maghemite, and ferrite spinel is considered the most probably assemblage
created by the diffusion of cations from the ferrite spinel crystal
structure. Higher temperatures, however, appear to favor the formation of
hematite rather than maghemite (11, 12, 13).

In some magnetic particles, merging of dendrites as growth proceeds is
thought to generate ferrite spinel crystals of octahedral habit. This type
of particle is observed to have good development of hematite lamellae
within a network of what appears to be coarsely dendritic ferrite spinel
(Fig. 5, 6). Oxidation along octahedral planes indicates that octahedra
develop from merging dendrites; these are also thought to produce curved
boundaries between ferrite spinel and the surrounding matrix.

Figure 6. Conversion of ferrite spinel
particle to hematite. Re-
flected light, oil immersion;
bar = 25 μm.

The dark matrix phase occurring between dendrites is considered to be
silicate glass because of its extremely
low reflectance, isotropic, transparent
optical properties and its spatial relationships to the reflective iron-
rich crystals. These spherical bodies
resemble ferrospheres observed by Lauf
(14). They differ in that the phase
between the magnetic crystals is not
mullite but a silicate glass. Ferrospheres are thought to be pseudomorphs
of whole framboids. These bodies lack the morphology of framboids.

Microscopy of the Magnetically Impoverished Fraction

Nonmagnetic fractions, because of their predominantly transparent
mineralogy, were examined in thin section and as loose crystal mounts
immersed in oils. It was seen that quartz, mullite, and hematite decreased
in abundance with particle size. Glassy, spherical particles were the

predominant type in all fractions. Crystals such as ferrite spinel and lime cannot be distinguished from dark or clear glassy particles. Some fractions were observed to contain many irregularly shaped particles (Fig. 7.

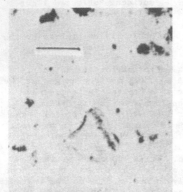

Figure 7. Detrital quartz crystal from fly ash. Transmitted light, oil immersion; bar = 25 µm.

Many of the particles can best be described in terms of the crystalline phase or phases which may constitute all or parts of them. The following is a description of the crystalline phases commonly found in nonmagnetic fly ashes examined in this study. Each crystalline phase is discussed with respect to identification and its relation to particle properties.

Quartz:

Quartz is distributed throughout nonmagnetic MA and KA fly ash, but, as determined from X-ray diffraction patterns, is concentrated in coarser size fractions. In thin section, quartz particles are clear, angular fragments showing clear evidence of conchoidal fracture (Fig. 7). Because the immersion medium used in these studies has an index of refraction of 1.515, the quartz particles are almost invisible in plane polarized light. Most, however, are quite coarsely crystalline and give good interference colors between crossed polarizers. Some of the quartz particles are monocrystalline and others are polycrystalline, but all show a similar morphology. The particles are typically rust stained and have spherical masses fused to them. Some quartz particles contain small opaque, black crystals resembling ferrite spinel. Such crystals, on the basis of previous studies of quartz inclusions, are considered to be ferrite spinel for the purposes of this study. At any rate, the black cubic inclusions crystallized in the quartz before it became part of the coal. Some of the quartz crystals are slightly rounded, but the general irregular shape, pronounced conchoidal fracture, and the coarse crystal size indicated that the quartz here described is not a product of crystallization in the furance. Watt and Thorne (15) report that silica from the breakdown of clay minerals may crystallize in the furnace as quartz; but this material is most probably detrital quartz that has been in the coal as quartz since the time of peat accumulation. Considering the viscosity of siliceous melts and the rapid quenching experienced by fly ash particles, we would be very surprised to find newly crystalline quartz in fly ashes.

Mullite:

Mullite is regarded as a disordered phase intermediate between the two better ordered structures, sillimanite and andalusite, with partial replacement of silicon by aluminum and a shift of some tetrahedral cations into open sites. Fe^{+3} and Ti may replace Al^{+3} in the mullite structure. Though it resembles quartz in thin section, mullite can be distinguished by higher index of refraction and pleochroism.

As observed in these specimens, mullite occurs in extremely fine-grained (<30 µm) crystals which form rounded, irregularly shaped, polycrystalline particles, the matrix of some aggregate type particles and rims surrounding opaque or glassy cores. Occasionally, spherical particles, as large as 15 µm, composed entirely of mullite are observed. The presence of mullite is considered due to the decomposition of clay minerals in the furnace.

Hematite:

A small amount of hematite occurs in the nonmagnetic ash. As mentioned above, hematite has weak magnetic susceptibility and occurs in the magnetic fraction only because it is intergrown with ferrite spinel. In transmitted light studies, the red coloration on very thin edges and sharp extinctions make the identification of hematite positive. The mineral also occurs as brown stains on other crystals and as intergrowths with ferrite spinel crystals that are fused to large nonmagnetic particles.

Lime:

Lime, optically isotropic, is difficult to distinguish microscopically from silicate glasses in these specimens. Because of this, it has probably not been recognized in some occurrences. X-ray diffraction patterns from whole ash samples before any separation had been made showed diffraction maxima thought to be characteristic of lime. Patterns from the magnetically impoverished samples showed these maxima with higher intensities. A sample with very intense maxima suspected of being caused by lime was immersed in water overnight, removed from water, dried and again subjected to the X-ray diffraction experiment. The maxima suspected of being caused by lime were not observed. Particles having very high indices of refraction were isolated and subjected to analysis by X-ray diffraction techniques. In all instances, the samples gave good powder patterns characteristic of lime. The most common appearance of lime in these ash specimens was as discrete spherical particles or flakes, that appeared to be fragments of cenospheres, of white sugary texture (Fig. 8). Microscopic observation of lime as discrete, unmixed particles supports similar conclusions drawn from density separations, solubilities and X-ray determinations. The fact that lime is unmixed with any other phase probably indicates that no other ions were available for reaction at the temperature where the parent clacite decomposed. It is known that the decomposition of calcite to CaO + CO_2 in the presence of elemental sulfur, or one of its oxides, in the conditions of furnace combustion results in the formation of oldhamite, CaS (16).

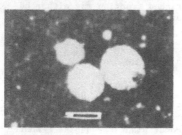

Figure 8. Lime particles. Transmitted light, oil immersion; bar = 120 µm.

Coal:

Some ash particles observed in the nonmagnetic fractions of KA and MA fly ashes were found to be unburned or partly burned coal. The uncombusted particles are angular and irregular in shape. These show layered coal structure viewed under low magnification; mounted and examined in vertically incident, reflected light, the macerals are readily identified. Partially combusted coal occurs as char spheres and fragments of char spheres. Some of these particles are dark colored, irregularly shaped, and vesicular or lacy in appearance (Fig. 9). Lightman and Street (17) and Ramsden (18) describe the appearance and development of this sort of particle during different stages of combustion.

28

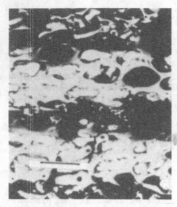

Figure 9. Incompletely combusted coal.
400x, reflected light, oil
immersion; bar = 25 μ.

Glassy spheres:

The concentration of glass spheres
ranges from about 10% in KA and MA 1 to about
95% in KA and MA 4. The glass varies in
color from clear to yellows, reds and grays
in transmitted light. Typically the sphere
surface is smooth and many contain bubble
inclusions that may have resulted from the
evolution of a gas (22, 56). Some of the
glass spheres have been broken, perhaps by
rapid expansion of contained gases or rapid
cooling.

CONCLUSIONS

As a result of the experiments and observations described above, we
feel it may be concluded that:

(1) It is not possible to make a very clean separation of magnetically
attracted and nonmagnetically attracted phases in one pass through the Roto-
Flux separator. (2) Magnetic attraction is related primarily to the ratio
of magnetic to nonmagnetic material making up the particle and secondarily
to the particle mass. (3) Concentration of phases in a magnetic separate
is controlled to some extent by the power setting of the separator.
(4) Bulk chemical content of the magnetic fraction is a function of
separator power setting and the number of susceptible particles in the fly
ash having a high or low ferrite spinel/glass composition. (5) Magnetic
particles are predominantly spheres consisting of three intergrown phases:
Ferrite spinel, hematite, and glass in varying proportions. (6) Hematite
observed in the magnetically enriched fraction results from either oxidation
of a ferrite spinel-rich spinel or polymorphic inversion from maghemite.
(7) Quartz observed in the magnetically enriched fraction may either be
entrapped in the magnetically attractive particles or may contain ferrite
spinel inclusions.

REFERENCES

1. Jarrige, A. 1970. Research in the area of fly ash. United States
    Bureau of Mines Information Circular 8488: 220-236.

2. Murtha, M. J. and Burnet, G. 1978. Power plant fly ash as a source
    of alumina. Presented at the Sixth Mineral Waste Utilization, Chicago,
    Illinois.

3. Murtha, M. J. and Burnet, G. 1978. New developments in the lime-soda
    sinter process for recovery of alumina from fly ash. Ames Laboratory
    USDOE and Department of Chemical Engineering, Iowa State University,
    paper No. IS-M-177.

4. Murtha, M. J. and Burnet, G. 1978. The magnetic fraction of coal fly ash: Its separation, properties and utilization. Proceedings of the IowaAcademy of Science 85: 10-13.

5. O'Gorman, J. V. and Walker, P. L. 1971. Mineral matter characteristics of some American coals. Fuel 50: 136-151.

6. O'Gorman, J. V. and Walker, P. L. 1973. Thermal behavior of mineral fractions separated from selected American coals. Fuel 52: 71-79.

7. Roy, N. K., Murtha, M. J., and Burnet, G. 1979. Use of the magnetic fraction of fly ash as a heavy medium material in coal washing. Proceedings of the 5th International Ash Utilization Symposium, Atlanta, Ga.

8. Roy, N. K., Murtha, M. J., and Burnet, G. 1978. Recovery of iron oxide from power plant fly ash by magnetic separation. International Conference on Industrial Applications of Magnetic Separators, Rindge, New Hampshire.

9. Minnick, J. I. 1961. The application of the Roto-Flux Magnetic Separator to pulverized coal fly ash. American Society of Mechanical Engineers. Paper No. 61-WA-313.

10. Colombu, U., Fagherazzi, G., Gazzarrini, F., Lanzavecchia, G., and Sironi, G. 1964. Mechanisms in the first stage of oxidation of magnetite. Nature 202: 175-176.

11. Columbu, U., Fagherazzi, G., Gazzarrini, F., Lanzavecchia, G., and Sironi, G. 1965. Magnetite oxidation: a proposed mechanism. Science 147: 1033.

12. Hurst, V. J. and Styron, R. W. 1976. Fly ash for use in the industrial extender market. AMAX Resource Recovery Systems, Inc., Smyrna, Georgia.

13. Keyser, T. R., Natusch, D. F. S., Evans, C. A., and Linton, R. W. 1978. Characterizing the surfaces of environmental particles. Environmental Science and Technology 12, No. 7: 768-773.

14. Lauf, R. J. Microstructures of Coal Fly Ash Particles. Ceramic Bulletin 61: 4, 487-490.

15. Watt, J. D. and Thorne, J. 1965. Composition and pozzolanic properties of pulverized fuel ashes I. Composition of fly ashes from some British power stations and properties of their component particles. Journal of Applied Chemistry 15: 585-594.

16. Biggs, D. L. and Lindsay, C. G. 1984. High temperature interactions among minerals occurring in coal. American Chemical Society, Division of Fuel Chemistry Symposium, Chemistry of mineral matter and ash in coal, In press.

17. Lightman, P. and Street, P. J. 1968. Microscopical examination of heat treated pulverized coal particles. Fuel 47: 1, 7-28.

18. Ramsden, A. R. 1968. Application of electron microscopy to the study formation. Journal of the Institute of Fuel: 451-454.

SCANNING ELECTRON MICROSCOPY AND X-RAY DIFFRACTION ANALYSIS OF VARIOUS SIZE FRACTIONS OF FLY ASH

R.C. JOSHI, G.S. NATT, R.L. DAY, and D.D. TILLEMAN
Department of Civil Engineering, The University of Calgary, Calgary, Alberta
Canada

(Received 29 November, 1984; accepted 25 February, 1985; Refereed)

ABSTRACT

X-ray diffraction and scanning electron microscopic analyses have been performed on the various Western Canadial fly ash fractions and on mixtures of cement and fly ash fractions. The ash mineralogy of hydrated and unhydrated fly ash samples was examined by x-ray diffraction. Results from a study of the morphology of ash particles etched in dilute hydrofluoric acid are also included. Scanning electron microscopy indicates that most of the reactive portion of the ash is on the particle surface. Also, as the fly ash particle size increases, the amount of crystalline SiO_2 increases but the quantity of calcium containing compounds decreases. Crystalline SiO_2 (the majority of which is present in the +75 μm fraction) does not take part in self-hardening reactions but does appear to take part in pozzolanic reactions.

INTRODUCTION

Fly ash has been used as a partial replacement for cement in concrete for almost five decades. Fly ashes produced by burning pulverized coals in suspension-fired power plants are classified as pozzolans. Specifications for use of ash as a replacement for cement in concrete place stringent restrictions on ash properties. For example, ASTM and CSA specify that no more than 34% by weight of a given ash be coarser than 45 microns (μm) if the ash is proposed for use as a pozzolan in concrete. More severe are British standards which restrict the amount of +45 μm fraction to 12%.

Two reasons have often been advanced for limiting the fraction coarser than 45 μm. First, some studies on bituminous coals indicate that the ashes containing large proportions of coarse material also contain large amounts of unburnt carbon [1,2]. Second, other studies suggest that the greater the proportion of +45 μm particles, the lower is the reactivity of the ash [3,5]. In general, it has been suggested that an increased amount of carbon particles as well as reduced reactivity makes the use of ash with a high proportion of coarse particles unattractive.

Some of the western fly ashes, particularly the sub-bituminous and lignite fly ashes, contain more than 34% by weight of +45 micron particles. Because of the standard specifications, suppliers of these ashes have to size classify their ashes, at a very high cost, to allow use of the ash with cement in concrete. However, studies by Joshi [6] indicate that classification of some of the western fly ashes may not be necessary. Most western fly ashes contain very little carbon and the proportion of particles larger than 45 μm does not seem to affect the reactivity of these ashes to any great extent.

Figure 1 shows results of the pozzolanic activity index (PAI) test on two Alberta ashes containing various proportions of +45 um material [5]. Although there is a trend for reduction in PAI as the proportion of +45 μm particle increases, the reduction is not as large as one might expect; there is no sudden drop in PAI when the amount of +45 um fraction exceeds 34%. Even the 100% +45 μm ash shows appreciable pozzolanicity.

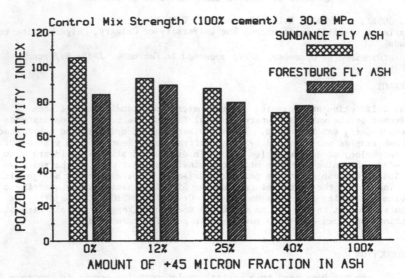

FIGURE 1. Pozzolanic Activity Index of Fly Ash Samples with
Various Proportions of +45 μm Fraction.

The role of particle size in determining the pozzolanicity of an ash is
still not certain. The majority of fly ashes contain very large proportions
of spherical particles of alumino-siliceous glassy material. It is important
to investigate the proortion and reactivity of glass in various size fractions
in order to understand pozzolanic reactivity.

In this paper, results of studies on one of the ashes noted in Figure 1
are presented. Mineralogical components and particle morphology of the
hydrated and unhydrated fly ash particles were determined. X-ray diffraction
and scanning electron microscopy (SEM) were used to study the reaction
products. Also, morphological and elemental analyses were performed on ash
particles in different size ranges and on the particles etched in 1%
hydrofluoric acid (to dissolve the glassy portion of the ash).

EXPERIMENTAL

Forestburg fly ash derived from sub-bituminous coal in Alberta, Canada
was selected for this study. Chemical, physical and pozzolanic properties of
the raw ash (as supplied from the plant) are given in Table I. This ash is
also known to possess cementitious properties. The raw ash was cut into four
different size fractions. A nest of sieves, numbers 60, 100, 200 and 325, was
used. Sample designations and general descriptions of the four samples are
given in Table II. Type 10 Portland cement and distilled water were used for
preparing pastes of cement and fly ash fractions.

Fly ash particles from various fractions (as detailed in Table II) were
examined in the SEM. X-ray diffraction and energy dispersive x-ray (EDX)
analyses were also run on the different samples.

TABLE I

Properties of Forestburg Fly Ash

CHEMICAL CHARACTERISTICS

Silica (SiO_2)	48.4%
Alumina (Al_2O_3)	23.5%
Iron Oxide (Fe_2O_3)	5.0%
Calcium Oxide (CaO)	17.0%
Titanium Oxide (TiO_2)	<1.0%
Magnesium Oxide (MgO)	0.5%
Sulphate (SO_3)	0.3%
Sodium Oxide (Na_2O)	3.2%
Potassium Oxide (K_2O)	1.1%
Loss on Ignition	0.4%

PHYSICAL CHARACTERISTICS

Magnetic Portion	<1.0%
Water Soluble Fraction	5.4%
Specific Surface Area (cm^2/g)	3690
Specific Gravity	2.04

Similar tests were conducted on moist samples of ash fractions cured for different times, as well as on pastes of equal weights of cement and fly ash. These samples were cured for 7 and 28 days at 20°C and 100% relative humidity before examination.

Examinations were also performed on samples A to D (Table II) after etching in 1% hydrofluoric acid. Particles from different size fractions were etched after they were mounted on stubs (in preparation for SEM). Epoxy was used to mount the specimens, and the stubs were lowered into the different solutions in a plastic beaker. The solutions were continuously agitated with a plastic coated magnetic stirrer. After specific times (10 minutes to 17 hours) the stubs were removed from the solutions, washed with distilled water, vacuum dried and coated with gold for SEM and EDX analysis.

A Phillip's W1139 diffractometer was used with a molybdenum target and a zirconium filter. The x-ray unit was operated at 20 mA and 40 kV. Specimens were scanned at a rate of 1 degree per minute. X-ray intensities were recorded on a strip-chart. A count rate of 500 per second and a time constant of 10 seconds were used.

Table II

Description of Samples Used in This Study

Sample	General Description	Size Fractions
A	finest	passing #325 mesh sieve (-45 μm)
B	intermediate	passing #200 mesh but retained on #325 mesh sieve (-75+45 μm)
C	coarse	passing #100 mesh but retained on #200 mesh sieve (-150+75 μm)
D	coarsest	passing #60 mesh but retained on #100 mesh sieve (-250+150 μm)

A Cambridge SEM-150 equipped with an energy dispersive x-ray analyzer (EDX) was used in all of the studies. EDX allowed elemental analyses of whole particles and areas of particles to be undertaken.

Specimens for SEM and EDX were prepared by mounting the ash samples onto copper stubs; colloidal silver was used as the adhesive. To eliminate or minimize charging effects all specimens were gold coated.

RESULTS AND DISCUSSION

X-Ray Diffraction

Unhydrated Ash. The diffraction traces of Sample A (all particles less then 45 microns) showed some predominant and many low intensity peaks and a hump. The peaks are believed to be representative of the crystalline phases of quartz, calcium hydroxide and some calcium silicates. A comparison of diffraction traces of the various ash fractions (samples A to D) indicated that although the quartz peaks increased only slightly in fractions B and C, they increased substantially in fraction D, the coarsest fraction. Another significant observation was that the peaks representative of the calcium compounds gradually decreased in size from the finest to the coarsest fraction (samples A to D).

The results indicate that the amount of calcium-bearing compounds decreases and the amount of quartz increases as the size of the fly ash particles increases. McCarthy et al. [7], on the other hand, found coarse fractions of fly ash richer in lime than finer fractions in a commercial Minnesota fly ash. At this stage it would appear that generalizations cannot be made about correlations betwen mineralogy and fly ash particle size.

A broad maximum representing glassy or amorphous material was present in all traces. No significant differences could be observed in the characteristics of this maximum among the different size fractions. This suggests that the amount of glass in the different fractions is approximately the same.

Hydrated Ash. In all samples, after 7 days of hydration, noticeable changes in the diffraction traces were observed. The calcium hydroxide peaks disappeared and the intensities of the peaks of 2.02 and 2.33 A decreased significantly. Reduction of these peaks seems to be representative of a reaction process which produces cementitious compounds; another study showed that samples cured 7 days or more developed significant compressive strengths [8]. In contrast, the intensities of quartz peaks were unchanged in all of the samples, thus suggesting that quartz did not take part in the self-hardening reactions.

In general, diffraction traces of 28 day and 7 day cured samples were similar, although the reductions in peak intensities at 2.02 and 2.33 A were more advanced.

Hydrated Fly Ash/Cement Mixtures. Compared among themselves, pastes made from cement and various size fractions of ash did not show significantly different x-ray traces. Crystalline SiO_2 peaks were still present even after 28 days hydration. The intensities of the peaks, however, were much lower than those of the unreacted ash and of the 28 day cured hydrated ash samples. This observation suggests that the combination of portland cement with ash may produce conditions which act as a catalyst for the reduction of crystalline silica.

SEM Observations and EDX Analyses.

Unetched Particles. An examination of SEM micrographs of unetched particles from size fractions A through D indicated that the majority of particles of Forestburg ash are spherical. However, the proportion of spherical particles decreased as the size fraction increased; sample C and the coarsest sample (D) showed large numbers of irregularly shaped particles, most of which were quartz with some unburnt carbon. Quartz particles were easily detectable by morphology as well as by their characteristic silicon peak.

Figures 2 and 3 show representative spherical particles and cenospheres filled with other particles, which comprised the majority of the finest fraction (A) and intermediate fraction (B). EDX analysis of the particle in Figure 2 showed high Ca and Si (comprising approximately 40% of the total) and about 20% Al. (Note that results have been normalized so that the three major compounds comprise 100%.) Particles of this type tended to have a rough surface. Other particles, such as shown in Figure 3, were richer in Si. The proportion of Si, Al and Ca was generally 63, 25 and 12% respectively. These particles had, for the most part, a smooth surface morphology.

Irregularly shaped particles, as shown in Figure 4, were nearly pure silica. These particles were most often found in the coarsest size fractions.

Etched Particles. EDX analysis of many particles, both etched and unetched, and collation of results for different size ranges suggested that a classification based upon the composition of the etched ash could be used to categorize particles into the following types:

1. Particles high in calcium (Figure 5). The proportions of Ca, Al, and Si in this type of particle were observed to be approximately 55, 30, and 15% respectively.
2. Particles high in silicon (Figure 6). These particles were often found covered by needle shaped crystals. Particles shown in Figure 6 contained 60% Si, 35% Al and 5% Ca.
3. Particles rich in aluminum (Figure 7). Such particles contain aluminum and silicon in the ratio of about 3 to 1. Little or no calcium was observed.

36

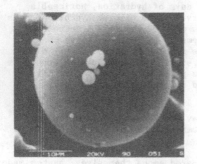

FIGURE 2
Unetched Ca & Si-Rich Particle

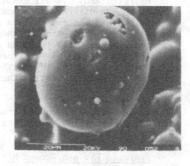

FIGURE 3
Unetched Si-Rich Particle

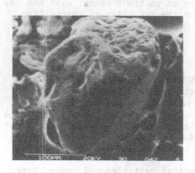

FIGURE 4
Silica Particle

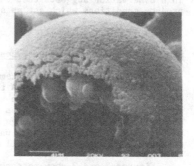

FIGURE 5
Particle Etched 10 Min.

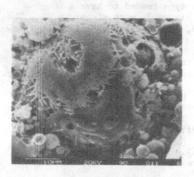

FIGURE 6
Particle Etched 10 Min.

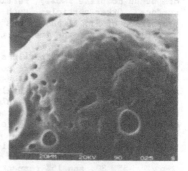

FIGURE 7
Particle Etched 17 hrs.

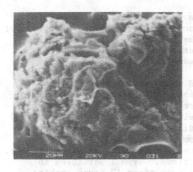

FIGURE 8
Particle Etched 17 Hrs.

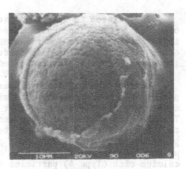

FIGURE 9
Surface Layer Etching

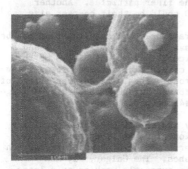

FIGURE 10
Hydration Between Particles
(14 day water cure, unetched samples)

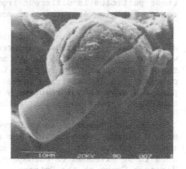

FIGURE 11
Elongated Particle

4. Irregularly shaped particles (Figure 8) showed only calcium when analyzed by EDX.

A significant amount of etching was observed in all four fractions after soaking in 1% HF. The proportion of particles which were etched after 10 minutes during this process increased as the size of the fraction decreased. This correlates with another observation mentioned above, that as the fraction size decreased the number of particles of Type 1 and 2 (particles rich in calcium and silicon) increased.

Silicon-rich particles showed extensive etching within a period of 10 minutes. When most of the fine particles were etched the proportion of Si was reduced compared to Ca and Al. Apparently there is a silica-rich coating on the particles, which was preferentially dissolved in 1% HF. An examination of Figure 9 tends to confirm that spherical particles are formed in layers of various substances of different chemical composition. Etching of the outer layer can be observed clearly in Figure 9. The micrograph in Figure 10 shows that the material from two spherical particles has dissolved and/or hydrated into a cementing compound joining the two particles. Observations of Figures 9 and 10 further suggest that the glassy or cementitious material in these spherical particles is concentrated on the outer surface.

In addition to spherical particles, some elongated particles were observed in the fine and intermediate fractions (samples A and B) (Figure 11); these particles were sometimes attached to the spherical particles, while other times they existed individually. EDX analysis of these elongated, irregular shaped grains indicated that they were very rich in Ca.

Most of the particles of samples A and B which were etched for long periods (17 hours) showed extensive dissolution. Many particles were completely dissolved; those that remained were of type 3 and 4 (see above). The skeleton of the particles generally contained Al and Si in the ratio of 3 to 1. Hulett et al. [9] also observed that silicon-rich particles were preferentially dissolved. Most particles from the coarser fractions (75-150 μm and 150-250 μm) were rich in silica. However, there were very few, if any, calcium-rich (Type 4) particles in these fractions. Type 1 particles in samples C and D were not completely dissolved even after 17 hours etching. This suggests that even the glassy particles, rich in calcium and silicon, in the coarse fractions are not as reactive as those in the finer fractions. It is quite possible that the thermal history, and thus resultant solubility, of these particles is different from those of the finer particles. Another reason for slow dissolving of samples C and D could be their coarser size.

On the basis of SEM, EDX and x-ray diffraction, the four fractions may be divided into two graphs. Apparently, the glassy particles of sample A and B which are finer than 75 μm are much more siliceous and soluble in 1% HF than the +75 μm material in samples C and D. The crystalline material seems to be concentrated in particles larger than 75 microns. Thus, the adverse effect of particle size on ash reactivity would be exhibited in ashes containing a large number of +75 μm particles, rather than the +45 μm particles.

Further research is required to quantify the effect of coarse particles and to examine more closely the coarse-fraction glass which appears to have reduced reactivity. Also, the experiments performed here on a particular ash need to be extended to encompass many more ashes. The Calgary Fly Ash Research Group at the University of Calgary is currently working on a large-scale fly ash project.

CONCLUSIONS

Based on the analysis of various size fractions of a particular fly ash, it is concluded that there are significant mineralogical and morphological differences among the various size fractions of an ash. The amount of calcium-bearing compounds decreased and quartz increased as the size fraction of the ash increased. Apparently the large amount of quartz present in the coarser fractions did not take part in self-hardening reactions, but did take part, to some extent, in pozzolanic reactions. Quartz particles were not affected when soaked for 17 hours in 1% HF acid. Glassy particles present in the ash were, for the most part, siliceous. Alumino-silicate particles were more soluble in HF than the calcium-silicate particles. Surfaces of most spherical particles seem to be more reactive than the non-spherical particles. Most of the nonreactive material appeared to be present in the ash fraction coarser than the 200 mesh (75 μm) sieve. The arbitrary restriction on the amount of particles greater than 45 μm does not seem to be valid for reactivity considerations.

ACKNOWLEDGEMENTS

Funds for this research were provided by Pozzolanic International (Alberta) Ltd. of Calgary, Canada, and by the Natural Sciences and Engineering Research Council of Canada.

REFERENCES

1. Watt, J.D., and Thorne, D.J., J. Appl. Chem., 15, 585-594 (1965).
2. Vincent, R.D., Mateos, M. and Davidson, D.T., Iowa Highway Research Board, Bulletin No. 26, pp. 141-174, (1961).
3. Davidson, D.T., Sheeler, J.B. and Delbridge, N.G. Jr., Iowa Highway Research Board, Bulletin No. 26, pp. 81-93 (1961).
4. Littlejohn, R.F., Inst. of Fuel Journal, 39, 59-67 (1966).
5. Day, R.L., Joshi, R.C., Langan, B.W. and Ward, M.A., Second Intnl. Conf. on Ash Technology and Marketing, London, England, Sept. 1984.
6. Joshi, R.C., Proc. 6th Intnl. Ash Utilization Symposium, Reno, Nevada, pp. 480-489, March 1982.
7. McCarthy, G.J., Swanson, K.D., Keller, L.P. and Blatter, W.C., Cem. Concr. Res., 14, [4], 471-478 (1984).
8. Joshi, R.C. and Lam, D., Research Rept. CE81-6, Dept. of Civil Eng., University of Calgary, July 1981, pp. 40.
9. Hulett, L.D., Weinberger, A.J., Ferguson, N.M., Northcutt, K.J. and Lyon, W.S., Electric Power Res. Inst., EA-1822, Research Project 1061, Final Report, May, 1981.

ELECTROKINETIC PHENOMENA AND SURFACE CHARACTERISTICS OF FLY ASH PARTICLES

R.I.A. MALEK AND D.M. ROY
Materials Research Laboratory, The Pennsylvania State University
University Park, PA 16802

(Received 29 November, 1984; Accepted 20 February, 1985; Refereed)

ABSTRACT

The zeta-potentials of two fly ashes were studied (high-calcium and low-calcium). It was found that they possess a point of charge reversal at pH = 10.5 to 12. The point of zero charge (low-calcium fly ash) was found to be at pH = 5. Furthermore, it shifted to more acidic values after the fly ash is aged in several calcium-containing solutions. The surficial changes that could happen when mixing fly ashes with cement and concrete were further evaluated by aging fly ashes in different solutions: $Ca(OH)_2$, $CaSO_4 \cdot 2H_2O$, NaOH and water solutions. Information from analyses for different ionic species in the solutions and characterization of the solid residues (XRD and SEM) was used in tentative explanations for the different behavior of the two types of fly ash in cementitious mixtures and concrete.

INTRODUCTION

Extensive work has been devoted to characterization of fly ash particles, both high- and low-calcium species [1–7], and their incorporation in cementitious mixtures and concrete [8]. The pozzolanic activity index required by ASTM C 618 to select the proper fly ash for incorporation in cement and concrete relies upon 28–day testing results, and it would be desirable to have additional chemical criteria [9–10]. The reaction between fly ash and lime has several applications in the field and use of low-calcium fly ashes in concrete seems satisfactory especially when sufficient time is available for these slow reactions and strength development to take place [9]. Therefore, it seems questionable to place so much reliance on the pozzolanic activity index, also since the fly ash continues to hydrate and the concrete properties change far beyond 28 days. Several investigators have reported significant calcium accumulation on the surface of fly ash particles during early hydration, with minimal interaction with the fly ash particles in cement and concrete [13–16].

This led us to re-examine the nature of interactions between fly ashes and different ionic species in concrete pore fluids. The surface characteristics and charge were also determined to provide better understanding of the nature of fly ash surfaces and the way they affect the hydration process. The zeta-potential of fly ash particles was studied by Roy et al. [11], who reported that fly ash particles need only minute amounts of superplasticizer to generate a strongly negative surface and enhance the workability of pastes. This could be attributed to the specific nature of fly ash particles. It is the major aim of this paper to examine the surface characteristics of different fly ash particles, their specific interactions with different ionic species, and the importance of these interactions with respect to the performance of fly ashes in cement and concrete.

EXPERIMENTAL

Materials and Methods

Two fly ashes were studied, high-calcium (air permeability [Blaine] surface area = 338.8 m^2/Kg) and low-calcium (Blaine = 276.6 m^2/Kg). Chemical analyses are given in Table I, expressed as oxides. X-ray diffraction identified (in addition to glass) the following mineralogical constituents: quartz, lime, periclase, mullite, ferrite spinel, alite, and anhydrite (for the high-calcium fly ash); and quartz, mullite, and ferrite spinel (for the low-calcium fly ash).

Leaching experiments were carried out by placing one-gram samples of the fly ash in plastic vials with 10 ml of four different solutions. Deionized water, saturated Ca(OH)$_2$, 1N NaOH, and saturated CaSO$_4$·2H$_2$O were placed separately on the fly ash particles. The plastic vials were sealed, shaken vigorously for 30 minutes, and placed immediately in an oven at 40°C. At designated times, 5 ml portions of the supernatant liquid were withdrawn from each vial for analysis. Precautions were taken to avoid withdrawal of suspended solid material using filter tipped syringes. The solutions were analyzed by DC plasma emission spectrometry (for cations) and by Dionex automated ion exchange chromatography (for anions). In some cases when a gel was found suspended in the supernatant liquid, part of the solution was vacuum dried and the resulting solid was analyzed. The solid materials remaining after withdrawal of the supernatant liquid were collected, washed and subjected to other experimental characterizations.

Table I. Chemical Analyses of Ashes, wt %.

Oxide	High-Lime (B15)	Low-Lime (B25)
SiO$_2$	40.60%	50.20%
Al$_2$O$_3$	17.40	27.00
TiO$_2$	21.24	1.40
Fe$_2$O$_3$	5.05	13.80
MgO	3.58	0.84
CaO	26.86	1.82
MnO	0.026	0.034
SrO	0.23	---
BaO	0.58	---
Na$_2$O	0.96	0.24
K$_2$O	0.72	2.45
P$_2$O$_5$	0.65	0.49
SO$_3$	1.40	1.36
CO$_2$	(0.17)*	---
L.O.I.**	0.47	0.48
Totals	99.77%	100.11%

*Not included in total, probably due to particle carbonation of CaO.
**Loss on Ignition, 900°C.

Electrokinetic Studies

The electrophoresis technique was followed to measure the electrophoretic mobility (cm^2/V·S) of the charged fly ash particles. A Zeta-Meter (ZM-77 manufactured by Zeta-Meter, Inc., 1720 First Avenue, New York, NY 10028) was used. Suspensions of 200 mg fly ash in 100 ml deionized water were used at room temperature. From the electrophoretic mobilities, zeta-potentials were calculated according to the relationship:

$$ZP = 4\pi\eta(T)/D(T) \times EM$$

where ZP = zeta-potential in electrostatic units, $\eta(T)$ = viscosity of the suspending liquid at temperature (T), D(T) = dielectric constant of the suspending liquid at temperature (T), and EM = electrophoretic mobilities at actual temperature (T).

When a voltage is imposed across the electrophoretic cell, frictional heat is generated by the moving ionic species, causing the temperature of the liquid to rise continuously. If the rate of temperature rise is too high, thermal turbulence of the suspended particles makes the measurement inaccurate. The rate of temperature rise depends on the specific conductance of the suspending medium as well as the applied voltage. The specific conductance of each suspension was measured prior to zeta-potential determination, and a maximum recommended voltage was applied, to minimize the temperature rise at a particular specific conductance value. The applied voltage, however, was found in some cases to require further reduction to maintain the particle velocity in the range ~0.5 micrometer per second in order to eliminate the 'human factor' in creating measurable errors while tracking particles. In spite of the above experimental precautions, a slight temperature rise could not be eliminated. The temperature of each suspension was measured before and after each zeta-potential determination and the average temperature was used to find the corresponding viscosity and dielectric constant of the suspending liquid. A temperature rise of 1 to 2°C was detected in the present measurements where the average velocity of up to 30 particles was determined over a period of time not exceeding one minute. An error in zeta-potential values of $\pm 2\%$ is thereby due to this temperature approximation.

The pH dependence of zeta-potential was also studied. For these measurements, the suspensions were buffered prior to zeta-potential determination using dilute HCl and NaOH solutions. Part of the buffered suspension was used for zeta potential determination and meanwhile, the pH of the remaining part was continuously monitored. After the zeta-potential determination, the pH of the liquid from the cell was measured and found to match that of the continuously-monitored suspension. In cases where low-calcium fly ash was used, insignificant pH drift was observed, whereas an appreciable drift was observed for suspensions of high-calcium fly ash. In the latter case, zeta-potential measurements were almost impossible. This limitation could be avoided by using very dilute suspensions, unfortunately beyond the detection limit of the equipment used in the present investigation. Thus, the zeta-potential values reported are restricted to those measured on low-calcium fly ash, postponing the evaluation of the high-calcium ash until further experimental development is carried out.

Residual Solid Phase Examination

The starting materials as well as the solid residues after separating the supernatant liquids were characterized by XRD and SEM. A Phillips 3600 automated diffraction unit equipped with CuKα source was used to obtain XRD patterns. The operating voltage was 40 KV at 30 mA current and step scans of 0.02 degrees with counting times of 0.30 seconds were used.

An ISI-DS-130 scanning electron microscope operating at 40 KV and equipped with Kevex energy-dispersive x-ray detector was used for characterizing particle morphology and chemistry.

RESULTS AND DISCUSSION

Leaching Experiment

Specific Ionic Interaction with Fly Ash Surfaces: Figure 1 shows the variations in total alkali concentration (Na$^+$ + K$^+$) in different suspensions with time. It is observed that, except in NaOH solution, the amounts of alkali metal ions dissolved from the low-calcium fly ash remained constant over the entire period of the test. This amount represents only 5.1% of the total sodium and potassium present in this fly ash. No aluminum nor iron were detected in those three solutions, nor did the calcium sulfate solution affect the solubility of the alkali-containing phases. Therefore, it was assumed that about 5% of the total alkali metal ions (Na$^+$ + K$^+$) are associated with soluble salts (probably sulfates) that are presumably

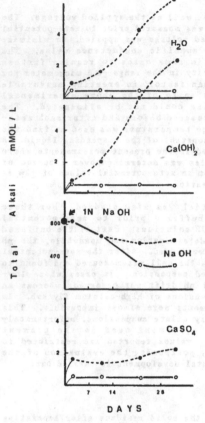

Fig. 1. Total alkali metal ions
$(Na^+ + K^+)$ concentrations
in different solutions,
H_2O, 1N NaOH, and saturated
$Ca(OH)_2$ and $CaSO_4$, vs.
time. (——) low-calcium;
(---) high-calcium fly ash.

present on the surface of low-calcium fly ash spheres, such as reported elsewhere [17]. The other 95% are incorporated in the relatively insoluble solid materials (glass) that form the bulk of the fly ash particles. The high-calcium fly ash behaves differently with respect to alkali metal ion solubility. In aqueous and saturated lime suspensions, the alkali metal ion concentrations increased with time indicating that most of the alkalis are probably incorporates in a more soluble glass, which leaches or hydrates steadily. By one month, 21%, 29%, and 8% of the total alkalis present in the solid high calcium fly ash were dissolved in water, saturated $Ca(OH)_2$ and saturated $CaSO_4 \cdot 2H_2O$, respectively. The smaller concentration of alkali in solution in the sulfate suspensions of high-calcium fly ash may be explained by an impermeable membrane formed early by precipitation of ettringite, minimizing further dissolution. Alternately, the alkalis may be incorporated in precipitation products.

Figures 2 and 3 give the concentrations of calcium and silica in different suspensions. Little or no Ca^{++} ions were detected in the 1N NaOH solution at all ages. The effect of alkali concentration on Ca^{++} concentration in hydroxide solutions has been shown to suppress Ca^{++} concentration with increasing alkali concentration [18,19]. The effect of sulfates on this Ca^{++}-alkali equilibrium was shown to increase the Ca^{++} ionic concentration at a given alkali concentration and pH [20].

Figure 2 shows that the Ca^{++} concentration in H_2O is initially higher for high-calcium than for low-calcium fly ash. The Ca^{++} concentration reaches the saturation level in two weeks and then drops. With the low SO_4^{--} concentration detected in aqueous solutions of high-calcium ash (0.3 m Mol/L at 30 days), the present results indicate the formation of insoluble phases with the Ca^{++} ion. The low silica concentration (Fig. 3) in aqueous suspensions of high-calcium ash further suggests that insoluble calcium silicate hydrate has been formed.

The Ca^{++} ion concentration in aqueous suspensions of low-calcium ash is relatively low, reflecting both its low concentration in the ash and the general insolubility of the ash, while the sulfate concentration is somewhat higher (8 m Mol/L SO_4^{--} at 30 days). The silicon concentration is considerably higher than that released from high-calcium ash. Evidence for

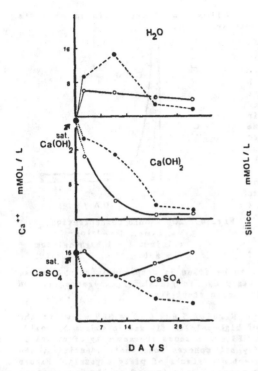

Fig. 2. Concentration of Ca^{++} dissolved from fly ash in different solutions vs. time. Ca^{++} dissolved in 1N NaOH is negligible at all ages. (——) low-calcium; (–––) high-calcium fly ash.

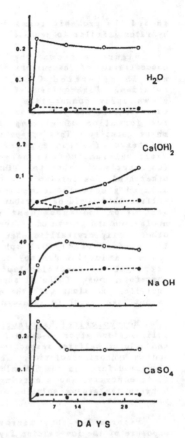

Fig. 3. Silica concentrations in different solutions vs. time. (——) low-calcium; (–––) high-calcium fly ash.

reaction of the Ca^{++} with low-calcium ash is seen from the Ca(OH)$_2$ suspensions, where the concentration drops steadily with time, while in CaSO$_4$ suspensions it drops a little and then increases. Thus Ca^{++} ions are apparently involved in some sort of interaction at the surface of fly ash spheres.

An XRD examination of the solid residue recovered showed the formation of calcium silicate and aluminate hydrates as well as portlandite [suspension in Ca(OH)$_2$] and ettringite [suspensions in CaSO$_4$·2H$_2$O] in high-calcium fly ash. There was little evidence of the formation of these compounds with low-calcium ash (even when a sample of unwashed solid residue was examined). This indicates that Ca^{++} ions are primarily adsorbed on the surface of low-calcium fly ash with a small extent of chemical reaction.

In NaOH solution, both fly ashes show high dissolution of silica (higher in the low-calcium ash) which increases with time up to 10 days. Aluminum and iron were detected in supernatant liquid indicating severe attack by NaOH on the bulk solid fly ash. XRD patterns of the solid residue showed the probable formation of sodium-substituted calcium silicate hydrates (similar to jennite) in both ash types.

Figure 4 shows the SO_4^{--} concentration of suspensions of fly ash in saturated $CaSO_4 \cdot 2H_2O$ solution. High-calcium fly ash shows early consumption of the sulfate. The XRD pattern indicated the formation of ettringite in major quantity. This suggests the existence of soluble phases to yield calcium and aluminate ions (to react with sulphate). On the other hand, low-calcium ash showed delayed gradual consumption of sulfate that could be attributed to adsorption and subsequent slow nucleation and growth of gypsum and other poorly crystalline phases on the surface of fly ash spheres. X-ray examination did not detect ettringite or monosulfoaluminate.

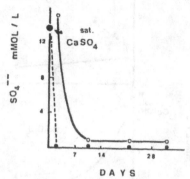

Fig. 4. SO_4^{--} ionic concentration vs. time. (——) low-calcium; (---) high-calcium fly ash.

Therefore, most of the Al appears to be fixed in a sparingly soluble phase (mullite, and along with the necessary Ca, in glass). Furthermore, the pH is not high enough to maintain ettringite stability.

Morphological Examination: Figures 5 and 6 show SEM views of the solid residue after 30-day aging of high-calcium fly ash in sodium hydroxide and calcium sulfate, respectively. Figure 5 shows the severity of attack in sodium hydroxide solution. The fly ash spheres lose their identities; the elongated forms in the SEM micrograph are stacks of platy crystals. Figure 6, in contrast, shows ettringite crystal rods formed by the interaction between the gypsum solution and dissolved calcium and aluminate ions from the fly ash.

Figures 7-9 show micrographs of the solid residues after 30 days exposure of the low-calcium fly ash to different solutions. The micrographs reveal the dissolution and precipitation features on the surface of fly ash spheres resulting from slow dissolution or hydrolysis.

Fig. 5. High-calcium fly ash aged in 1N NaOH, 30d. Fly ash spheres are hardly identifiable. Severity of reaction exceeds the surface layer. EDXA shows Na-substituted Ca silicate and aluminosilicate hydrates.

Fig. 6. High-calcium ash aged 30d in $CaSO_4$ solution. 2-2μ, elongated ettringite crystals accumulate around ash spheres. Ash surface dissolution is apparent.

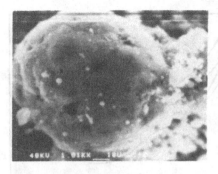
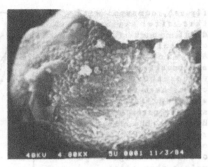

Fig. 7. Low-calcium ash aged 30d in
water. Some features of
dissolution and/or
accumulation of solid
matter on surface are
shown.

Fig. 8. Low-calcium ash aged in 1N
NaOH for 30d. Dissolution
features with crystalline
remnants exposed. EDXA
suggest hematite or
magnetite.

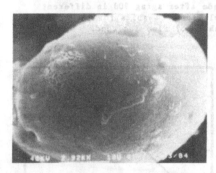

Fig. 9(a). Low-calcium ash aged 30d
in saturated $CaSO_4 \cdot 2H_2O$
solution. The surface
seems covered with
crystalline material.

Fig. 9(b). Surface of particle in
Fig. 9(a). EDX analysis
indicates crystals are
gypsum.

Surface Charges and Zeta-Potentials

As discussed, the Ca^{++} reacts in a complex manner with species
dissolved from the high-calcium ash and abundant new products are formed,
while it is physically adsorbed on the surface of the low-calcium ash, where
it can also act as nucleation sites for the formation of calcium-containing
crystals. To further investigate the effects introduced by adsorbing or
reacting species on surface characteristics, zeta-potentials were measured
for the original ash as well as the solid retained after aging for 30 days
in different solutions. Deionized water was used as a suspending medium in
all cases studied. It was found that suspensions of high-calcium ash
possess very high conductances even before aging that made measurements of
zeta-potentials extremely difficult using the available Zeta-Meter. Thus,
the current discussion and results are limited to the measurements for low-
calcium fly ash. This class of ash is of special interest, in that respect,
since it is the one that shows specific physical adsorption. Figure 10
shows the pH dependence of the zeta-potential for the original low-calcium

48

fly ash, compared with that after aging in different Ca^{++}-containing solutions. All are strongly electronegative at pH values above 7. The curves have the common shape and features of silicate particles [12]. However, the point of zero charge (PZC) of the original low-calcium ash occurs at higher pH values than silica and many silicate particles. This is presumably because aluminosilicate and ferrite phases are incorporated with the siliceous glass, possibly on the surface. Aging the fly ash particles in different solutions produces an appreciable effect on the surface charge and particle mobility. The PZC shifts to more acidic values compared with the original fly ash. In fact, the entire curves shift their positions extending over the whole range of pH indicating an extensive modification of the particle surface either by hydrolysis or precipitation of layers of new composition. At pH > 10, mobilities decrease in all cases because of the coagulation of particles and measurements in that range become difficult. Extrapolation could give a point of charge reversal at about pH = 12. This relative

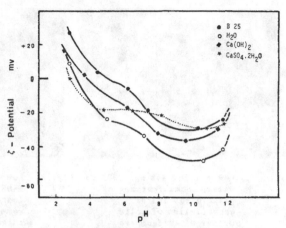

Fig. 10. The pH dependence of zeta potentials for low-calcium ash. Measurements made after aging 30d in different solutions. Results for the original ash (B25) also included.

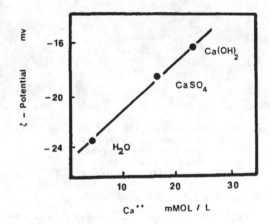

Fig. 11. Dependency of zeta potential of low-calcium ash aged in different solutions on Ca^{++} concentration of the aging solution.

behavior may, to some extent, reflect the early stage effects which influence the dispersion of fly ash particles during mixing with cement pastes where they are in contact with Ca^{2+}-containing solutions. The shift of PZC to more acidic values indicates a removal of positive ions to increase the electronegativity of the surface. In pure water the fly ash surface appears to be leached to achieve this effect. The drop in particle mobility (zeta-potential) at pH = 5 (the pH attained for suspensions in water without buffering) is proportional to the initial Ca^{++} ion concentration in the solution; e.g., H_2O aging solutions produced the highest zeta potential. A linear relationship is shown in Fig. 11. It appears that the low-calcium ash has a preferential tendency to adsorb Ca^{++} ions.

CONCLUSIONS

1. In high-calcium fly ash, there is a tendency for reactions involving calcium ions, silica and probably some other ions on the surface, forming hydrated compounds. When incorporated in cement pastes, low-calcium fly ash furnishes higher silica content to the solution, thus, reacting with the Ca^{++} ions released from cement to form bonding compounds. Provided that enough time is allowed, appreciable strength gain is the consequence.

2. Alkalis are released continuously into solution from the high-calcium ash. $Ca(OH)_2$ enhances their dissolution, whereas $CaSO_4$ retards it. When gypsum is present in sufficient amount, ettringite will form in the early stages.

3. Low-calcium fly ash seems to have a preferential tendency to adsorb Ca^{++} ions and their compounds. Five percent of the total alkali ion content is present as soluble salts on the surface of fly ash spheres; the rest is incorporated in an insoluble form in the solid material (glass) of the spheres.

4. Long-term leaching will cause a shift in the PZC to more acidic values; more negative zeta-potentials are the result of aging in solutions with the lowest Ca^{++} concentration.

REFERENCES

[1] L.D. Hulett, A.J. Weinberger, N.M. Ferguson, K.J. Northcutt, and W.S. Lyon, EA-1822 Research Project 1061, EPRI Final Report (1981).

[2] L.D. Hulett, A.J. Weinberger, K.J. Northcutt, and N.M. Ferguson, Science 210, 1356 (1980).

[3] R.J. Lauf, ORNL TM-7663, 71 pp. (1981).

[4] R.J. Lauf, Amer. Ceram. Soc. Bull. 61 (4) (1982).

[5] G.J. McCarthy, K.D. Swanson, L.P. Keller and W.C. Blatter, Cem. Concr. Res. 14 (4), 471-478 (1984).

[6] G.J. McCarthy et al., Cem.Concr. Res. 14 (4), 479-484 (1984).

[7] R.J. Stevenson, Cem. Concr. Res. 14 (4), 485-490 (1984).

[8] S. Diamond, in Proc., Symposium N, Effect of Fly Ash Incorporation in Cement and Concrete, Ed. S. Diamond, p. 12. Materials Research Society, University Park, PA (1981).

[9] S. Diamond, Cem. Concr. Res. 14 (4), 455-462 (1984).

[10] S. Schlorholtz, T. Demirel, and J.M. Pitt, Cem. Concr. Res. 14 (4), 499-504 (1984), and references cited therein.

[11] D.M. Roy, M. Daimon, and K. Asaga, Proc. 7th Intl. Congr. Chem. Cement,Vol. II, II-242, Editions Septima, Paris (1981).

[12] R.B. Ottewill, Phil. Trans. R. Soc. A310, 67-78, London (1983).

[13] Fajun Wei, M.W. Grutzeck, and D.M. Roy, Cem. Concr. Res. 15, 174 (1985).

[14] K. Ogawa, H. Uchikawa, and K. Takemoto, Cem. Concr. Res. 10, 863 (1980).

[15] I. Jawed and J. Skalny, in Proc., Symposium N, Effects of Fly Ash Incorporation in Cement and Concrete, Ed. S. Diamond. Materials Research Society, University Park, PA (1981).

[16] Jun-Yuan He, B.E. Scheetz, and D.M. Roy, Cem. Concr. Res. 14, 505 (1984).

[17] D.M. Roy, K. Luke, and S. Diamond, Characterization of Fly Ash and its Reactions in Concrete (this volume).

[18] R. Hedin, Proc. Swed. Cem. Concr. Res. Inst., No. 3 (1945).
[19] N. Fratini, Annali. Appl. 39, 616 (1949).
[20] W. Rechenberg and S. Sprung, Cem. Concr. Res. 13, 119 (1983).

TECHNICAL NOTE ON THE DETERMINATION
OF FREE LIME (CaO) IN FLY ASH

SCOTT SCHLORHOLTZ AND TURGUT DEMIREL
Iowa State University, Department of Civil Engineering, Ames, IA 50011

(Received 8 February, 1985; Communicated by G.J. McCarthy)

Many fly ashes contain free lime (CaO) and periclase (MgO) [1,2]. These two compounds, when present in excessive amounts, are known to cause soundness problems in portland cement [3,4]. Recent work [5] has indicated that the autoclave expansion of portland cement-fly ash pastes is related to the concentration of CaO and MgO in a given paste, free lime typically being more detrimental than periclase. The purpose of this technical note is to briefly discuss two methods that are currently available for determining the free lime content of fly ash, and to suggest a supplement to the autoclave test (described in ASTM C 151). The major drawback of the autoclave test is that it requires approximately two days to complete and therefore it would be helpful to have a quick chemical test that could be used to indicate the soundness properties of a given fly ash.

Presently there are two major techniques used to measure free lime in fly ash: 1) a volumetric method of chemical analysis [4,6] and 2) quantitative x-ray diffraction techniques [1].

The volumetric method is not new. It has been used for years to monitor the free lime content of portland cement [4,6]. Briefly, this method uses a solvent to extract the free lime from the bulk material (fly ash, portland cement, quicklime, etc.), a filtration procedure to separate the solvent plus extracted lime from the bulk material and a quantitative step consisting of a titration of the filtrate with a standard acid. Unfortunately, the volumetric method extracts both CaO and Ca(OH)2 and sometimes other soluble compounds of calcium, and this is the major drawback of the test. CaO is detrimental to soundness while Ca(OH)2 is not. Also, choosing a suitable extraction solvent is not a simple task because fly ashes exhibit large variability in chemical (and compound) composition; hence, what may work for one fly ash may not be adequate for another fly ash. Obviously water must not be used as a solvent because many fly ashes contain significant amounts of soluble calcium containing compounds that could potentially be mistaken for free lime.

Anhydrous ethylene glycol [4,6] and ethyl acetoacetate-isobutyl alcohol or glycerin-ethanol [6] solvents have also been used as extractants. The ethylene glycol method described by Javellana and Jawed [7] appears very promising; experimentation in our laboratory has shown that the method is quick (extraction times of 5 minutes at temperatures above 85°C) and test results are in good agreement with those obtained by quantitative x-ray diffraction. The major problem with the test is that extended extraction times can cause dissolution of calcium containing compounds other than free lime, this problem can be eliminated (or at least minimized) by proper use of specific analytical procedures that are calibrated with appropriate fly ash standards. The method is applicable only to "fresh" fly ashes which contain only unhydrated CaO.

Quantitative x-ray diffraction (QXRD) techniques can successfully be used to determine the concentration of crystalline compounds in fly ash [1]. The most appealing aspect of QXRD is the fact that it is a structure sensitive analytical technique; hence, calcium oxide can be measured independently of calcium hydroxide. Thus, QXRD is not restricted to the analysis of

52

"fresh" fly ashes. It can also be used to monitor the hydration rate and formation of hydration products in fly ash-water-portland cement systems. Cost is the major factor restricting the use of QXRD. Capital investment for a modern diffractometer and the cost of hiring trained technical personnel to run and maintain the instrument make it economically inaccessible to most testing laboratories.

In summary, two methods, wet chemistry and quantitative x-ray diffraction, that are applicable to the determination of free lime (CaO) in fly ashes have been discussed. Both methods have their own strengths and weaknesses. The wet chemical method is weak in the area of accuracy because the method is influenced by all of the soluble calcium containing compounds present in fly ash. Speed, simplicity and cost effectiveness are the major attributes of the wet chemical method. The x-ray diffraction method is weak in the area of cost effectiveness but very strong in the area of basic research.

Fly ash is a waste product and we cannot expect it to be subjected to expensive (i.e. time consuming or exotic) testing since the testing would increase the cost of the fly ash and effectively reduce its utilization. Yet, we must assure the quality of fly ash concrete at a cost equivalent to (or less than) that of normal portland cement concrete. More effort at clarifying which test specifications really have significance in governing the quality of fly ash is needed. Preparation of a series of fly ash standard reference materials with known phase as well as chemical composition would be of tremendous help in standardizing many of the chemical analysis techniques that have appeared over the years.

REFERENCES

1. M. L. Mings, S. Schlorholtz, J. M. Pitt, and T. Demirel, Transportation Research Record 941, 5-11 (1983).
2. G. J. McCarthy, K. D. Swanson, L. P. Keller, and W. C. Butler, Cement and Concrete Research, 14, 471-478 (1984).
3. H. F. Gonnerman, W. Lerch, and T. M. Whiteside, PCA Research Laboratory Bulletin 45 (June 1953).
4. F. M. Lea, The Chemistry of Cement and Concrete, 366-367 (1971).
5. S. Schlorholtz and T. Demirel, this symposium.
6. ASTM, 1981 Annual Book of Standards, Part 13 (1981).
7. M. P. Javellana and I. Jawed, Cement and Concrete Research, 12, 399-403 (1982).

CHARACTERIZATION OF CRYSTALLINE PHASES IN FLY ASH
BY MICROFOCUS RAMAN SPECTROSCOPY

BARRY E. SCHEETZ AND WILLIAM B. WHITE
Materials Research Laboratory, The Pennsylvania State University,
University Park, PA 16802

(Received 3 January, 1985; Accepted 6 February, 1985; Refereed)

ABSTRACT

The laser Raman microprobe was used to interrogate individual fly ash particles as small as 1 μm diameter and record "fingerprint" Raman spectra from both crystalline and non-crystalline components of the fly ash. When compared to reference patterns of known crystalline phases, the Raman spectra can be used to identify crystalline phases and can give some structural information on other phases. Furthermore, because this method characterizes fly ash particles on an individual basis, correlations to both color and morphology of the particles can be made.

INTRODUCTION

Fly ash produced from the combustion of coals serves as an extender to cement and in some cases as a reactive cementitious component. The chemical reactivity of fly ash depends on the phase composition of the crystalline components as well as the composition of the glassy components. Characterization of fly ash has routinely relied on both bulk chemical analyses and powder x-ray diffraction methods. Both of these techniques characterize bulk samples, integrating the results from individual particles to produce the data. Phase characterization by x-ray diffraction is often limited because the small amount of crystalline component present in the fly ash may be masked by the x-ray scattering of the glassy component. In some occasions there is only a profusion of overlapping low intensity lines [1]. A previous report described the use of microfocus Raman spectroscopy to characterize the glassy phases present in fly ash [2]. This paper discusses the crystalline phases and some of the problems attendent on their characterization.

MICROFOCUS RAMAN SPECTROSCOPY

The concept of combining the light microscope with the Raman spectrometer came independently from Dhamelincourt and his colleagues in France [3,4] and Rosasco and his colleagues at the National Bureau of Standards [5,6]. The laser source is directed through a beam splitter which directs the beam downward through the microscope optics. The beam emerges from the objective lens and is brought to a focus on the microscope stage. In principle, the limit on resolving power is the diffraction limit for focusing coherent visible light, often quoted as 1 μm. Practical limits depend on the particular microscope optics, and on the internal scattering of the sample. Certainly spatial resolutions of a few μm can be achieved easily. Raman scattered radiation from the sample returns along a nearly parallel beam path through the microscope optics and, on reaching the beam splitter, is directed to the entrance slit of the double monochromator. Beyond this point, the optical system is that of a conventional laser-excited Raman spectrometer with double monochromator to provide good stray light discrimination.

The spectra reported in this paper were measured on an Instruments SA Ramanor U-1000 spectrometer. The detector was a cooled RCA C-31034

photomultiplier with pulse-counting electronics. Individual pulses were fed
to a Columbia Prism microcomputer and stored on disc. Spectral displays
could then be arranged as required including time-averaging and multiple-
scan averaging. This data manipulation capability is particularly useful
for fly ash samples which are intrinsically rather weak Raman scatterers.
All spectra presented in this paper are taken directly from the plotter.
The laser source was a Spectra-Physics model 164 argon ion laser. Input beam
power from the 514.5 nm green line was held to 200 mW. The complex optics of
the microfocus instruments are not efficient and effective power at the
sample is 5-10% of the input power. However, because of the small beam diameter, power density is very high and for this reason one must be wary of
changes in the sample from beam heating. No sample preparation was required.
Fly ash was spread in a thin layer on a microscope slide. Particles to be
measured were selected from the image on the microscope projection screen,
and these particles were brought to the focal point by translating the stage.

FLY ASH SAMPLES AND MINERALOGY

Fly ash minerals arise from the inorganic components in the coal as modified by thermal reactions during their brief passage through the combustion
chamber. Minerals which have been identified in fly ash are listed in Table
I arranged according to the expected quality of Raman signal. Four fly ash
samples were examined in this investigation. These were a low lime fly ash
from Pennsylvania Power and Light Co. [7], a high lime fly ash from Dowell
[7], a lignite ash [8], and a North Dakota fly ash [9]. In this reconnais-
sance study, randomly selected particles from each fly ash were measured.
There was no attempt at statistical sampling so that no intercomparison between the fly ash samples can be made at this time.

RAMAN SPECTRA

Quartz

As might be expected, quartz is the most common crystalline phase in fly
ash. Figure 1 shows the spectra of three quartz grains. The top spectrum is
of a well-crystallized quartz, the middle spectrum is more typical of what is
usually observed, and the bottom spectrum is of a grain with some crystal
quartz embedded in a glassy particle. The Raman scattering from the glass is
very weak but appears as the broad band near 1100 cm^{-1}.

Table I. Minerals Identified from Fly Ash.

Good Raman Signal	Weak Raman Signal	No First Order Spectrum
Anhydrite	Mullite	CaO
Gypsum	Gehlenite	MgO
Thenardite	Hematite	
Aphthitalite	Magnetite	
K_2SO_4	Ferrite spinels	
Quartz		
Ca_3SiO_5		
Ca_2SiO_4		
$Ca_3Al_2O_6$		
Feldspar		

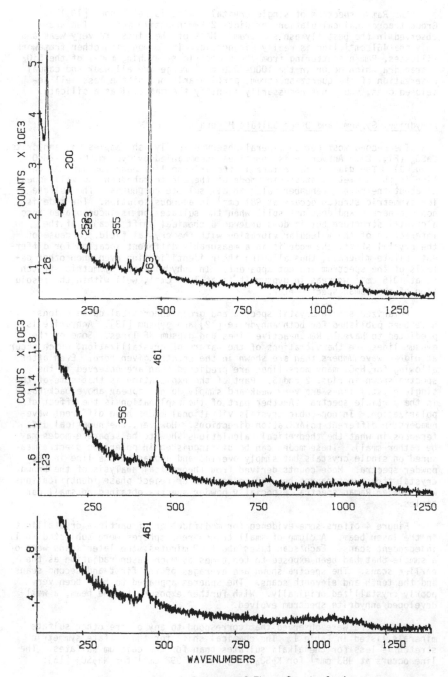

Figure 1. Raman Spectra of Three Quartz Grains

The Raman spectrum of single crystal quartz is well known [10,11]. Group theoretical calculations predict 12 Raman-active modes. Ten are observed in the best fly ash spectrum. Most of the lines are very weak and only the 461 cm^{-1} line is really diagnostic. In common with other framework silicates, Raman scattering from the symmetric stretching modes of the SiO_4 tetrahedra which occur in the 1000-1200 cm^{-1} range are all weak and casual observation of the quartz spectrum, particularly one of the less well developed ones, would not necessarily identify the material as a silicate.

Anhydrite, Gypsum, and Other Sulfate Minerals

The second most common mineral observed in fly ash samples was anhydrite, $CaSO_4$ (Fig. 2). Anhydrite is sometimes accompanied by gypsum, $CaSO_4 \cdot 2H_2O$ (Fig. 3). The diagnostic feature is the intense line near 1000 cm^{-1}. This line is the symmetric stretching mode of the SO_4 tetrahedron and will appear at about the same wavenumber value in all sulfate compounds. The sulfate ion symmetric stretch occurs at 981 cm^{-1} in aqueous solution. The mode is non-degenerate and does not split when the sulfate ion is incorporated into a crystal structure but it does undergo a chemical shift because of the interaction of the molecular vibration with the crystal field. Because of the crystal shift, the mode is in a measurably different location for different sulfate minerals, thus allowing their identification even when other details of the spectrum are not apparent. In anhydrite the symmetric stretch is at 1016 cm^{-1} whereas in gypsum it is at 1006 cm^{-1}, well within the resolution of the Raman spectrometer.

Polarized single crystal spectra and group theoretical calculations have been published for both anhydrite [12] and gypsum [13]. Anhydrite is predicted to have 18 Raman-active lines and gypsum 36 lines. Some of the gypsum lines are the vibrations of the waters of crystallization, which occur at higher wavenumbers than are shown in the spectra given here. Even after allowing for H_2O, many more lines are predicted than are observed in the spectra shown in Figs. 2 and 3. Part of the explanation is that some of the single crystal lines are very weak and simply do not appear above background in the particle spectra. Another part of the explanation is the effect of polarization. In non-cubic crystals vibrational modes have different wavenumbers in different polarization directions. However, the numerical differences in what the theoretical calculations show to be separate modes may be rather small. These modes can be distinguished in polarized spectra measured on single crystals but simply overlap and appear as one line in the powder spectra. Mode counts derived from theoretical analysis of the known crystal structures cannot be used to confirm or reject phase identifications with powder Raman spectra, especially powder spectra obtained on small particles.

Figure 4 offers some evidence for modification of particle crystallinity in the laser beam. A clump of small transparent spheres were subjected to 11 independent scans. Each scan takes about 70 minutes; the later scans were of a sample that had been exposed to ten times as much laser radiation as the earlier scans. The spectra shown are averages of the first and second scans and the tenth and eleventh scans. The spheres appeared to have been very poorly crystallized originally. With further exposure to the beam, a well developed anhydrite spectrum evolved.

No spectra were observed that correspond to any of the other sulfate minerals listed in Table I. The chemical shift for the sulfate symmetric stretch is less for the alkali sulfates than for the calcium sulfates. The line occurs at 983 cm^{-1} for K_2SO_4 [14] and at 992 cm^{-1} for Na_2SO_4 [15].

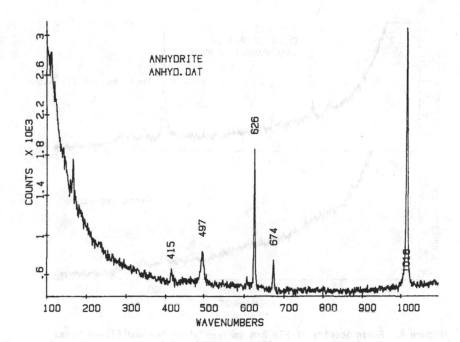

Figure 2. Raman Spectrum of Anhydrite, $CaSO_4$

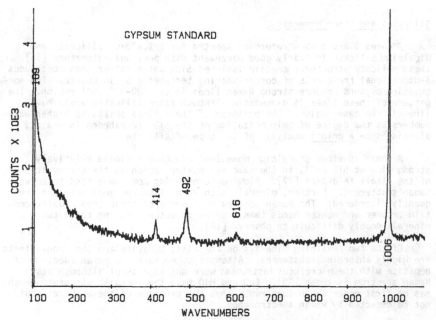

Figure 3. Raman spectrum of Gypsum, $CaSO_4 \cdot 2H_2O$

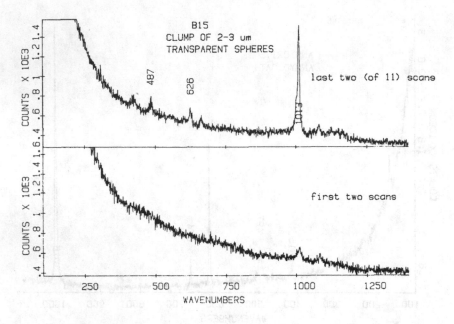

Figure 4. Raman Spectra of Fly Ash Spheres after Two and Eleven Scans
Showing the Recrystallization of Anhydrite Due to Beam Heating

Silicates and Other Minerals

Figures 5 and 6 show reference spectra for tricalcium silicates and
dicalcium silicate in fairly good agreement with previous literature [16].
These silicate structures contain isolated SiO_4 units rather than continuous
3-dimensional frameworks of corner-sharing tetrahedra as in quartz. The non-
bridging oxygens produce strong Raman lines in the 800-900 cm^{-1} region. The
pattern of these lines is diagnostic although other silicates would have
lines in the same region. The position of these lines shifts to higher wavenumbers as the degree of polymerization of the SiO_4 tetrahedra increases,
allowing some a priori analysis of the type of silicate.

A Raman spectrum of calcium monoaluminate gave a single relatively
strong line at 518 cm^{-1}, in the same wavenumber region as the strongest bands
of the alkali feldspars [17]. Aluminum-containing compounds tend to be weak
Raman scatterers. Further, minerals such as mullite and gehlenite are frequently disordered. The Raman spectra of disordered structures usually contain broader and weaker bands than ordered structures making these two
minerals doubly difficult to observe [18].

Of the two remaining minerals on the list, hematite and the iron-spinels
are highly absorbing substances. Attempts to measure the Raman spectrum of
hematite with the microfocus instrument were not successful although its
Raman spectrum is known [19]. CaO and MgO have the rocksalt structure which
has no first order Raman spectrum. Thus the presence of these phases would
not be detected by Raman spectroscopy.

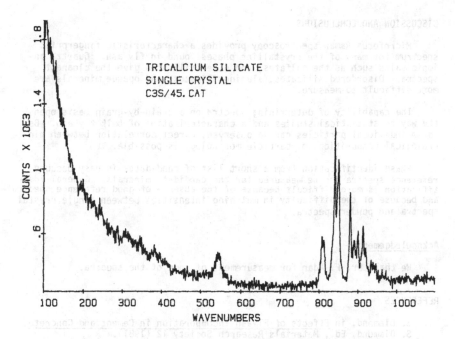

Figure 5. Raman Spectrum of Tricalcium Silicate, Ca₃SiO₅

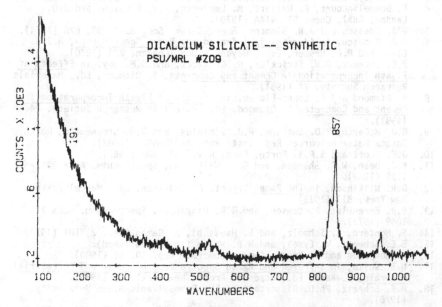

Figure 6. Raman Spectrum of Dicalcium Silicate, Ca₂SiO₄

DISCUSSION AND CONCLUSIONS

Microfocus Raman spectroscopy provides a characteristic fingerprint spectrum for many of the crystalline phases found in fly ash. Quartz, anionic salts such as the sulfates, and other silicates give the cleanest spectra. Disordered silicates, aluminosilicates, and opaque minerals are more difficult to measure.

The capability of determining spectra on a grain-by-grain basis opens the way to statistical studies and a characterization of bulk fly ash. Because individual particles can be observed, direct correlation between mineralogical composition and particle morphology is possible.

Phase identification from a short list of candidates is easy because reference spectra can be measured for the candidate minerals. General identification is more difficult because of the absence of good reference spectra and because of the difficulty in matching intensities between single crystal spectra and powder spectra.

Acknowledgements

We thank Dr. F. Adar for measurement of some of the spectra.

REFERENCES

1. S. Diamond, in Effects of Flyash Incorporation in Cement and Concrete, S. Diamond, Ed., Materials Research Society 12 (1981).
2. B.E. Scheetz, W.B. White, and F. Adar, Proc. Conf. on Characterization, Alfred University, Plenum Press (in press).
3. M. Delhaye and P. Dhamelincourt, J. Raman Spectros. 3, 33 (1975).
4. P. Dhamelincourt, F. Wallart, M. Leclercq, A.T. N'Guyen, and D.O. Landon, Anal. Chem. 51, 414A (1979).
5. G.J. Rosasco and J.H. Simmons, Amer. Ceram. Soc. Bull. 53, 626 (1974).
6. G.J. Rosasco, in Advances in Infrared and Raman Spectroscopy, R.J.H. Clark and R.E. Hester, Eds., Hayden & Don, London, 223 (1980).
7. B.E. Scheetz, D.W. Strickler, M. Grutzeck, and D.M. Roy, in Effects of Flyash Incorporation in Cement and Concrete, S. Diamond, Ed., Materials Research Society, 24 (1981).
8. S. Diamond and F. Lopez-Flores, in Effects of Flyash Incorporation in Cement and Concrete, S. Diamond, Ed., Materials Research Society, 34 (1981).
9. G.J. McCarthy, K.D. Swanson, P.J. Schields, and G.H. Groenewold, North Dakota Water Resources Res. Inst. Rpt. A-078-NDAK (1983).
10. J.F. Scott and S.P.S. Porto, Phys. Rev. 161, 903 (1967).
11. K.J. Dean, W.F. Sherman, and G.R. Wilkinson, Spectrochim. Acta 38A, 1105 (1982).
12. G.R. Wilkinson, in The Raman Effect, A. Anderson, Ed., Marcel Dekker, New York, 811 (1973).
13. B.J. Berenblut, P. Dawson, and G.R. Wilkinson, Spectrochim. Acta 27A, 1849 (1971).
14. S. Montero, R. Schmölz, and S. Haussühl, J. Raman Spec. 2, 101 (1974).
15. B.E. Scheetz, W. Eysel, and W.B. White (to be published).
16. M. Conjeaud and H. Boyer, Cement Concrete Res. 10, 61 (1980).
17. W.B. White, in Infrared and Raman Spectroscopy of Lunar and Terrestrial Minerals, C. Karr, Ed., Academic Press, New York, 325 (1975).
18. B.E. Scheetz, Ph.D. Dissertation, The Pennsylvania State University (1976).
19. I.R. Beattie and T.R. Gilson, J. Chem. Soc. (London) A5, 980 (1970).

CHARACTERIZATION OF CATALYZED DEVITRIFICATION IN QUENCHED FLY ASH MELTS

SUBHASH H. RISBUD
Department of Ceramic Engineering and Materials Research Laboratory,
University of Illinois at Urbana-Champaign, Urbana, IL   61801

(Received 11 February, 1985; Communicated by G.J. McCarthy)

Coal combustion produces enormous quantities of residual ash often called bottom ash or fly ash. The fly ash component contains lightweight cenospheres giving the ash a fluffy character. Fly ash is captured in the coal combustion process by air pollution control devices as the gases exit the stack. Fly ash compositions are usually highly siliceous consisting mainly of the oxides of silica, alumina, calcia, and iron oxides; minor constituents such as MgO, alkali oxides, TiO_2 etc. are also almost invariably present in quantities of ~0.5 to 3 wt%. Two important aspects of crystallization of fly ash melts and glasses relate to the prevention of boiler slagging [1] and, from a waste utilization point of view, to the development of new products using fly ash as a raw material[2-4]. Ash devitrification on cooling of the melt results in friable material that does not stick to boiler walls as easily as glassy slag [5]. From another standpoint, crystallization of glassy ash slag to a fine grained equiaxed microstructure is considered a desirable glass-ceramic body for thermomechanical reasons [6].

The present work is concerned with the characterization of the crystallization behavior of several fly ash compositions under different melt quenching conditions. The extent of crystallization in fly ash melts and glasses with nucleating additives was investigated via characterization of the quenched materials by DTA and TEM/STEM. In both low iron and high iron fly ash melts selected additives induced crystal nucleation at quenching rates of ~1000°C/sec to 1500°C/sec.

Three different fly ash compositions were used in this study: (i) a sub-bituminous ash typical of the combustion of coals in the western U.S.; (ii) a bituminous ash commonly resulting from the burning of coals found in the central and eastern U.S.; and (iii) an Illinois coal fly ash similar in composition to the bituminous ash. Fly ash powders were wet mixed with zircon, TiO_2, SiC, $MgAl_2O_4$ and a number of other additives all of which were powders -325 mesh or finer. The well blended mixture was dried by calcining at ~250°C to drive off the volatiles. Approximately 2 to 3 gms of additive modified ash was packed into a high alumina crucible. The alumina crucible was suspended by a Pt wire in a vertical tube furnace set at a temperature of ~1550°C. Melts were formed by holding at funace temperature for ~2 to 15 hours. Rapid quenching was achieved by dropping the alumina crucible into water thus cooling the melt at a rate estimated to be ~1000 to 1500°C/sec.

Melting endotherms in DTA heating traces and bulk melting in alumina crucibles were used to obtain liquidus temperatures for the

ashes while cooling curves using the DTA gave indications of
solidification transitions. The DTA melting behavior (in air) of
subbituminous ash suggested a liquidus temperature, T_L, of $\approx 1190°C$ in
comparison with a T_L of $\approx 1325°C$ for the bituminous ash. This increased
liquidus temperature for the bituminous ash is presumably due to its
higher Al_2O_3 and Fe_2O_3 content. Upon cooling, the molten ashes
solidified to a glassy mass. DTA cooling curves indicated no
crystallization peaks and electron diffraction in the TEM confirmed the
amorphous nature of the solidified material. DTA cooling curves of
ashes modified by nucleating agents catalyzed crystallization.
Crystallization in the subbituminous ash was nucleated by alumina,
zircon, AlN and SiC. The eastern bituminous ash appeared to be more
resistant to crystal nucleation although the AlN and SiC additives did
successfully cause the elimination of melt supercooling.

Additive modified sub-bituminous ash was fast quenched and the
quenched materials extracted from near the center of the alumina
crucible were floated on electron microscope grids for diffraction.
Extensive crystal growth had apparently not occurred during the fast
quench since no sharp crystalline peaks were observed in routine x-ray
diffraction. However, crystal nucleation could be detected by electron
diffraction patterns that suggested microcrystallinity in several
additive modified quenched ash melts. Nitride additives were specially
effective in catalyzing crystal nucleation during rapid quenching.

ACKNOWLEDGEMENT: Useful discussion of these results with D. F. Mahoney
are acknowledged.

REFERENCES

1. W. C. James and H. G. Fischer, J. Inst. of Fuel, 40, 170 (1967).

2. B. Butterworth, Trans. Brit. Ceramic Soc., 5, 33 (1954).

3. A. G. Pincus, Glass Industry, 53, 32 (1972).

4. W. Hinz and F. G. Wihsmann, Silikatechnik, 16, 110 (1965).

5. D. F. Mahoney, A. E. Kober, and S. H. Risbud, U. S. Patent
    4,372,227 (Feb. 8, 1983).

6. E. J. DeGuire and S. H. Risbud, J. of Materials Science, 19, 1760
    (1984).

PART II

Utilization of Fly Ash

RETARDATION EFFECTS IN THE HYDRATION OF CEMENT-FLY ASH PASTES

MICHAEL W. GRUTZECK, WEI FAJUN* AND DELLA M. ROY
  Materials Research Laboratory, The Pennsylvania State University, University
  Park, PA 16802
*Permanent Address--Wuhan Institute of Building Materials, Hubei, China

(Received 29 November, 1984; Accepted 28 January, 1985; Refereed)

ABSTRACT

The hydration of high-calcium and low-calcium fly ash-cement mixtures was
investigated to determine the effect of fly ash upon the hydration of a Type
I portland cement, and to determine the associated mechanisms of hydration.
When blended with portland cement, both fly ashes retarded the early hydration
process, the high-Ca more so than the low-Ca. Analyses of solution compositions and calorimetric (heat of hydration) measurements were made. The
retardation and hydration effects are discussed in terms of solution composition data and solid phase characterization. The hydration effects were
interpreted and compared with the results of previous work.

INTRODUCTION

In a recent paper, Wei et al. [1] observed that a high-calcium and a
low-calcium fly ash both retarded the hydration of Portland cement. It
was proposed that the fly ash acted both as a source of soluble alumina as
well as a nucleation and growth site for ettringite. Ettringite (AFt) was
observed to form early on the high-Ca fly ash surfaces. It was concluded
that the presence and condition of the fly ash surface, acting as a "calcium
sink," tended to depress the concentration of calcium ions in solution,
thereby preventing the active precipitation of $Ca(OH)_2$ and thus prolonging
the induction period. As a consequence, the maximum in the second hydration
peak was retarded when fly ash containing pastes were compared to neat cement
paste. These observations are in conflict with those of Ogawa et al. [2]
and Kawada and Nemoto [3] but very much in agreement with those of Jawed and
Skalny [4], Cabrera and Plowman [5] and Plowman and Cabrera [6]. A second
observation was that the more soluble high-Ca fly ash tended to retard the
hydration to a greater extent than the less soluble low-Ca fly ash.

The present paper is a continuation of the above work. Chemical data
are presented which show the effects of fly ash upon portland-cement solution
chemistry. The cement and fly ash-cement blends in question, formulated
with a water/solid (w/s) ratio of 10/1, were allowed to hydrate in sealed
plastic bottles at 38°C. By definition, the study of mixtures which contain
a relatively high proportion of water (w/s > 1), which act more like solutions than pastes, which are carried out in sealed plastic bottles, and
whose solutions are sampled primarily for chemical reasons are considered
"bottle-hydration studies." For sake of continuity, in addition to the
chemical data presented in this paper, a summary of some of the earlier work
by the same authors [1] is also presented.

EXPERIMENTAL METHODS

Earlier Work

Two fly ashes, a high-Ca and a low-Ca, were mixed with a Type I portland
cement and were hydrated in a Seebeck-type isothermal calorimeter maintained

at 38°C for 24 hours. The chemical composition and physical properties of the starting materials are given in Table I. The calorimeters were loaded with two-gram samples and mixed with 0.8 mL of injected deionized water. The rate of heat evolution of a Type I neat cement paste was compared to 60:40 (by weight) blends of cement and fly ash. The calorimeter studies led to the observation that the ashes retarded the second peak (i.e. prolonged the induction period). Once this had been established, six samples of each of the three mixtures (a total of 18 samples) were mixed with deionized water (w/s = 0.4) and cured at 38°C. At predetermined times, samples of each of the three pastes were washed, gently ground under acetone (to stop hydration) and dried in a vacuum oven at 38°C for approximately three hours.

TABLE I. CHEMICAL COMPOSITION AND PHYSICAL PROPERTIES OF STARTING MATERIALS.

	TYPE I* CEMENT	HIGH-CA FLY ASH	LOW-CA FLY ASH
SiO_2	20.84	36.8	47.6
Al_2O_3	4.10	18.1	25.7
TiO_2	0.27	1.55	1.16
Fe_2O_3	2.90	6.21	17.3
MgO	4.05	4.74	0.98
CaO	63.84	27.83	4.44
MnO	0.21	0.039	0.036
SrO	0.05	0.30	---
BaO	0.02	0.60	---
Na_2O	0.07	1.22	0.69
K_2O	0.74	0.41	1.59
P_2O_5	0.21	0.94	0.32
SO_3	2.67	1.90	---
LOI	1.32	0.55	2.76
TOTAL	101.29%	101.00%	102.58%
BET (m^2/g)	0.89	1.06	4.03

*PSU IDENTIFICATION NUMBERS I-10, B-62 AND B-42 RESPECTIVELY. HIGH-CA = ASTM CLASS C, LOW-CA = ASTM CLASS F.

The quantities of calcium hydroxide and non-evaporable water present in these hydrated cement and fly ash-cement blends were determined by thermogravimetric analysis (TGA) in a nitrogen atmosphere. The weight loss of the fixed-age cement paste was measured during a heating cycle up to 950°C, with temperature raised by 10°C/minute. Non-evaporable water was estimated by taking the weight loss between the temperature interval of 100°C and 950°C, and calcium hydroxide was evaluated by taking the weight loss between the temperature interval of 450°C and 510°C.

Present Work

Twenty-four hour "bottle-hydration studies" of the three mixtures described above were carried out at 38°C in sealed plastic bottles. The mixtures had a water/solid ratio of 10/1. A total of 21 samples (seven of each type) were prepared, each containing three grams of solid and 30 mL of liquid. The resulting slurries were allowed to hydrate and at predetermined times one of each of the three mixtures was sampled by vacuum filtering the suspension. The clear solution was acidified with HCl. The concentrations of ions in solution were determined with an SMI III DC plasma atomic emission spectrophotometer and a Dionex Model 2000i ion chromatograph. Because the experiments were of limited duration, the bottles sealed and the filtering process rapid, it was assumed that the affects of carbonation were negligible.

RESULTS AND DISCUSSION

Earlier Work

Figure 1 shows the rate of heat evolution of pure cement and fly ash-cement mixtures hydrated in deionized water at 38°C. The data show that both the high- and low-Ca fly ash retarded the cement hydration, with the high-lime fly-ash causing greater retardation. The data indicate that the addition of fly ash to the cement retarded the time to reach the second (major) exothermic peak and that the rate of heat evolution at the second peak was diminished relative to that of the cement, by the addition of

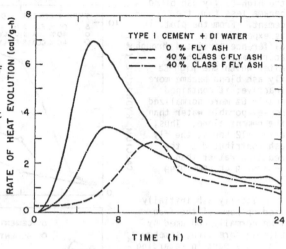

Figure 1. Rate of heat evolution of pure cement and fly ash-cement blends.

either high- or low-Ca fly ash (6.78, 5.03 and 5.88 cal/g·h, respectively, normalized to 100% portland cement).

The calcium hydroxide content of pure cement and fly ash (40)-cement (60) blends hydrated at 38°C and having a w/s ratio of 0.4 were determined by thermogravimetric analyses. The data obtained using the 450°-510°C temperature interval are presented in Figure 2. All pastes contained approximately 1% $Ca(OH)_2$ through three hours of hydration (Figure 2A). At that time $Ca(OH)_2$ formation increased dramatically in the pure cement. Even considering the dilution effect of the fly ash upon the cement [see Figure 2B, in which the percentage of $Ca(OH)_2$ in the neat cement has been subtracted from the percentage of $Ca(OH)_2$ (normalized to 100% cement) in the fly ash-cement blend] the fly ash cement pastes developed $Ca(OH)_2$ [or consumed $Ca(OH)_2$ released from cement hydration] at a much slower rate. Surprisingly, the high-Ca fly ash cement paste contained the least calcium hydroxide. By 24 hours, however, the cement, and the normalized high- and low-Ca pastes contained approximately the same (normalized) amount of $Ca(OH)_2$.

Figure 3 depicts the non-evaporable water content (derived from the 100° to 950°C weight loss) of pure cement, and fly ash (40)-cement (60) blends hydrated for 24 hours at 38°C. The 100-950°C weight losses were about the same for pure cement and high-Ca fly ash-cement pastes through three hours (Figure 3A). The values for the low-Ca fly ash-cement were relatively high, reflecting a high carbon content (note the high LOI in Table I for the low-Ca fly ash), and should be corrected for this. After three hours, the non-evaporable water in the cement increased relatively rapidly, whereas the fly ash blends gained relatively little non-evaporable water through six hours. By 24 hours, however, all three pastes had about the same degree of reaction; all contained about the same amount of non-evaporable water. Again, the normalized differences in non-evaporable water between the high-Ca fly ash blend minus its pure cement counterpart (see Figure 3B) revealed a familiar pattern of behavior. The pure cement and the high-Ca fly ash blend were about equally reactive up to three hours. At three hours the reactivity of the high-Ca fly ash blend became less than the cement. From the plot, it is expected that the difference persisted through approximately 12 hours after which the high-Ca fly ash blend became more reactive; it contained nearly 6% more normalized non-evaporable water than the cement alone. Thus, after ~12 hours, the fly ash contributed to the reaction rather than retarding or remaining inert.

The fly ash initially tied up some of the free lime normally consumed by $Ca(OH)_2$ formation, somehow playing a part in retarding the reaction. Scanning electron micrographs taken of the paste samples at three and eight hours established that the fly ash was a source of soluble alumina as well as a site for nucleation and growth of ettringite crystals.

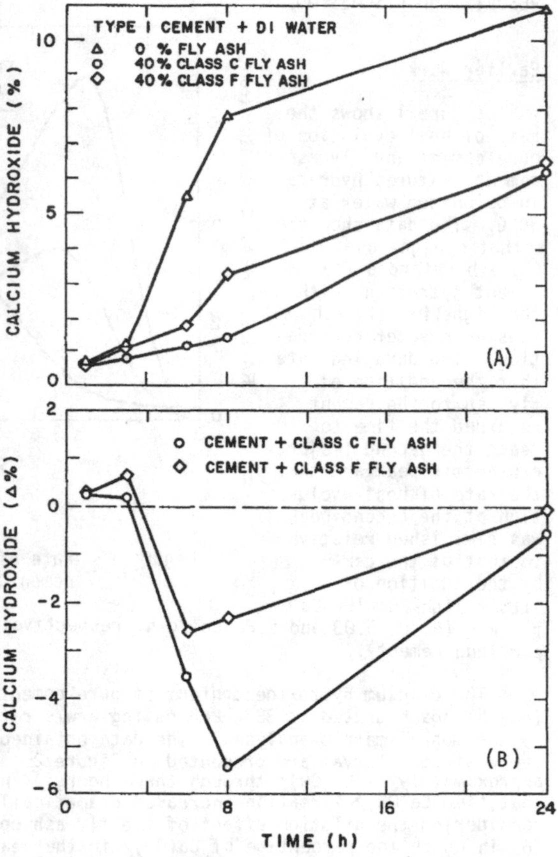

Present Work

In an attempt to correlate the above observations with solution chemistry, liquid compositions of both pure cement and fly ash (40)-cement (60) blends "bottle hydrated" at a w/s ratio of 10/1 were studied. Results are given as a function of time for calcium, sulfate, aluminum and silicon in Figure 4 and for potassium and sodium in Figure 5.

Figure 2.   (A) Quantity of $Ca(OH)_2$ in neat cement and fly ash-cement blends. (B) $Ca(OH)_2$ in normalized fly ash-cement blends minus $Ca(OH)_2$ in neat cement.

The chemical data illustrate a number of things. The calcium and sulfate concentrations (Figure 4), although initially quite variable, tended to converge at 24 hours, perhaps reaching some "equilibrium" value. For cement, calcium (view A) increased rapidly through one hour, then it leveled off, staying at approximately 1700 ppm until three hours, after which it increases through six hours and then decreased rapidly, approaching ~1000 ppm by 24 hours [~$Ca(OH)_2$ saturation]. The calcium content of the low-Ca fly ash-cement blend decreased from an initially higher value, leveling off at approximately 1100-1000 ppm through four hours, after which it increased through eight hours, followed by a decrease to approximately 1000 ppm Ca^{2+} by 24 hours.

It is suggested that the period of "level" concentration which spans the one- to three-hour range in the pure cement and which similarly spans one to four hours in the high-Ca fly ash-cement blend may be responsible for a build up of a Ca-rich layer on the clinker in the pure cement, and the fly ash and clinker in the fly ash-cement blends [1].

For all mixtures, sulfate (view B) drops in concentration over the 24-hour span of the experiment, and converges at about 100 ppm SO_4. During the first six hours, sulfate values for the fly ash blends were below those of the neat cement. Similar data have been reported by He et al. [7]. The concentrations of sulfate were far below the equilibrium values for gypsum solubility. Since abundant ettringite was forming, these values reflect the lower solubility of sulfate from ettringite [8].

At 24 hours the aluminum and silicon concentrations (Figure 4, views C and D) in solution from the fly ash blends were lower than those from the neat cement. However, at earlier times, the

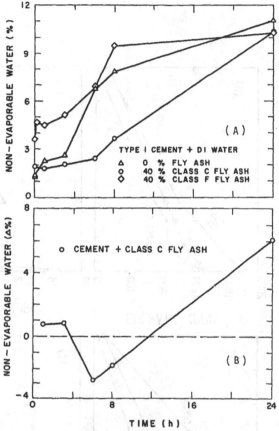

Figure 3. (A) Non-evaporable water in neat cement and fly ash-cement blends. (B) Non-evaporable water in normalized fly ash-cement blends minus non-evaporable water in neat cement.

behavior of both ions in solution was quite similar. Both fly ash blends showed a sharp spike in both aluminum and silicon concentrations over the first hour. Both blends had significantly higher concentrations of silicon and aluminum in solution than the pure cement. The pure cement generally did not have a spike in concentration of these ions. The concentrations of both silicon and aluminum were higher in the blends during the first five hours, after which they fell below the cement concentrations. The solution chemistry was different during the first five hours. At first the fly ash tended to supply silicon and aluminum to the solution which took part in the formation of ettringite, C-S-H, and perhaps some silica-gel as well.

The alkali ions were somewhat different in behavior. The potassium (view A) released from the cement was greater than 350 ppm, apparently from a soluble salt. With the addition of 40 weight percent fly ash (of either kind) the potassium content was lowered by almost exactly 40 percent,

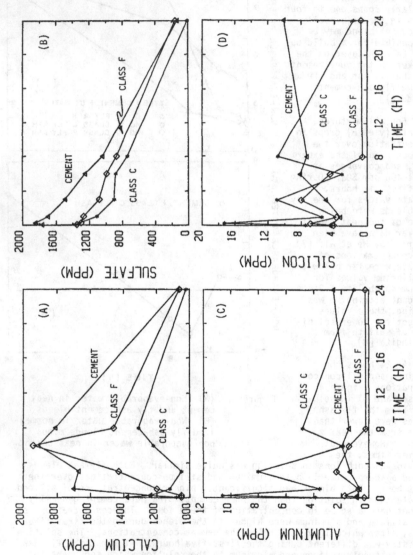

Figure 4. Concentration of ions in solution for "bottle-hydration studies" of neat cement (△) and fly ash-cement blends [*40 wt% high-Ca (C) fly ash and ◇ 40 wt% low-Ca (F) fly ash.] A=Ca, B=SO4, C=Al, D=Si.

suggesting a simple dilution effect, the cement supplying all the potassium. The levels of potassium remained constant for about six hours, which signifies a non-participatory role in the Stage I hydration and Stage II induction periods.

Whereas the potassium levels were fairly constant, the sodium (view B) behaved rather differently. The low-Ca fly ash contributed much more sodium to the system than did either the cement or the high-Ca fly ash. The sodium release patterns from the high-Ca blend and Type I cement were nearly identical which suggests that the sodium salts were present in nearly the same proportions, in both soluble and more slowly released forms. In the same fashion, the low-Ca fly ash must contain a greater percentage of soluble salts even though it had the lowest overall sodium content (Table I). Once more, the behavior of sodium suggested a buildup, rather than a lowering of solution concentration during the first 24 hours.

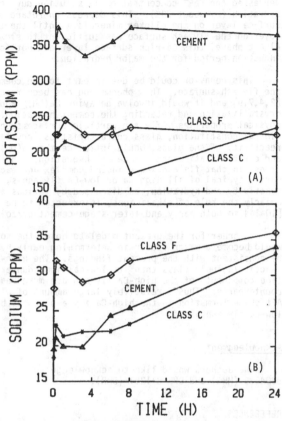

Figure 5. Alkali ions in solution for "bottle-hydration studies" of neat cement (Δ) and fly ash-cement blends (*40 wt% high-Ca fly ash and ◇ 40 wt% low-Ca fly ash). A=K, B=Na.

These data suggested that the alkali ions do not participate in forming insoluble reaction products during the first 24 hours of hydration. In a paste situation, this would imply that alkalis remain concentrated in the pore solutions, at least during the first 24 hours.

SUMMARY AND CONCLUSIONS

The chemical data are internally consistent and do not contradict previously suggested hydration models with respect to the retardation phenomena. In the instances examined, retardation is related to the presence and condition of the fly ash surfaces. It was suggested that the fly-ash surface acts somewhat like a calcium-sink. Calcium and sulfate in solution were removed by the abundant aluminum associated with fly ashes as an AFt phase preferentially formed on the surface of the fly ash. This depressed the Ca^{2+} concentration in solution during the first six hours of hydration (by about 30%). This depression retarded the formation of a Ca-rich surface layer on the clinker minerals. Until the thickness of the Ca-rich layer on the fly ash surface was sufficient to slow the formation of additional AFt phase, this Ca-rich surface layer did not form, resulting in a longer induction period for the major hydration.

This behavior could be due in part to the chemisorption of Ca^{2+} ions on the fly ash surface. This phenomenon has been reported by other authors [2,4,7,9] and it would involve delaying $Ca(OH)_2$ and C-S-H nucleation and crystallization and retarding the cement hydration. The chemical and physical properties of fly ash particle surfaces such as chemical composition, mineral constitution, glass characteristics, crystal defects, and the reactivity of the glass, would influence the $Ca(OH)_2$ and C-S-H nucleation and crystallization and the cement hydration. There are further qualifications, in that fly ashes are indeed complex and each sample is not necessarily typical of all high-Ca or low-Ca fly ashes, e.g., with respect to reactivity and hydration, as discussed by various authors [10,11]; particularly the role of alkali ions, is expected to be an important modifier [10-14] to both early and later stage cement-pozzolan (fly ash) hydration.

In order for the present model to hold, the solubility of the fly ash would become a major factor in determining early hydration. This seems to be consistent with the present findings. The high-calcium fly ash has a more reactive/soluble glass than its low-calcium counterpart. Also the crystalline components of the high-Ca fly ash are more reactive than their low-Ca counterparts. Thus a relatively large amount of aluminum is available for AFt phase formation in the high-Ca fly ash cement blend, compared to the low-Ca fly ash blend.

Acknowledgment

The authors would like to acknowledge the support of NSF Grants CEE-8304076 (MWG) and CPE-811281 (DMR).

REFERENCES

1. Wei Fajun, M.W. Grutzeck and D.M. Roy, Cem. Concr. Res. 15, 174 (1985).
2. K. Ogawa, H. Uchikawa and K. Takemoto, Cem. Concr. Res. 10, 863 (1980).
3. N. Kawada and A. Nemoto, Sement Gijutsu Nempo. 22, 124 (1968).
4. I. Jawed and J. Skalny, Effects of Fly Ash Incorporation in Cement and Concrete, Ed. S. Diamond, p. 60, Proc. M.R.S. Meeting, Boston, MA (1981).
5. J.G. Cabrera and C. Plowman, Proc. 7th Int. Cong. Chem. Cem. 3, IV-85 Paris (1980).
6. C. Plowman and J.G. Cabrera, Cem. Concr. Res. 14, 238 (1984).
7. Jun-Yuan He, B.E. Scheetz and D.M. Roy, Cem. Concr. Res. 14, 505 (1984).
8. F.E. Jones, J. Phys. Chem. 49, 344 (1945).
9. Zhao-Qi Wu and J.F. Young, J. Am. Ceram. Soc. 67 (1), 48 (1984).
10. G.M. Idorn and K.R. Henriksen, Cem. Concr. Res. 14, 463 (1984).
11. S. Diamond, Effects of Fly Ash Incorporation in Cement and Concrete, Ed. S. Diamond, p. 12, Proc. M.R.S. Meeting, Boston, MA (1981).
12. R.I.A. Malek and D.M. Roy, Proc. 6th Intl. Conf. on Alkalis in Concretes, Research and Practice, G.M. Idorn, Steen Rostam, Eds., p. 223, Danish Concrete Association, Copenhagen (1983).
13. W.T. Bakker, Ed., Workshop Proceedings: Research and Development Needs for the Use of Fly Ash in Cement and Concretes, EPRI CS-2616-SR, pp. 2-1 and 5-1, Electric Power Research Institute, Palo Alto, CA (1982).
14. D.M. Roy and G.M. Idorn, ACI Journal 79-43, 444 (1982).

REACTION PRODUCTS IN FLY ASH CONCRETE

MARK D. BAKER[*] AND JOAKIM G. LAGUROS[**]
* The Dolese Company, Oklahoma City, Oklahoma
** The University of Oklahoma, Norman, Oklahoma

(Received 24 October, 1984; accepted 28 February, 1985, Refereed)

ABSTRACT

The setting and strength gaining process of PC concrete containing Class C high lime fly ash were related to the reaction products identified using XRD and SEM. Four fly ash concrete mixes (20, 30, 40, and 50 percent replacement of cement by fly ash) and similar paste mixes were compared to control mixes for curing periods up to one year. Setting time and early compressive strength were adversely affected by the addition of fly ash. Beyond one week all of the fly ash concrete mixes gained strength at a faster rate than the corresponding control mixes. XRD studies suggest that the retardation mechanism may be associated with the high levels of ettringite formed early in the hydration process and its conversion to monosulfoaluminate. A decrease in the level of calcium hydroxide, typical of pozzolanic activity, was not in evidence. SEM micrographs of fly ash spheres in concrete at the various stages of hydration reveal an intricate crystal framework. A simple heat of hydration test is presented which helps explain the strength gains observed.

INTRODUCTION

Over the past decade, the availability of and the necessity to use Class C high lime fly ashes in the western United States have steadily increased. Unfortunately, the chemical and physical properties of these ashes are sufficiently different [1] from those of the Class F fly ashes, common in the eastern regions of the United States, to render much of the past research and experience with Class F ashes not entirely applicable to the Class C fly ashes. Thus, basic research into the effect of Class C fly ashes on concrete is needed.

This study was undertaken to evaluate the differences between the reaction products of PC concrete and fly ash concrete in which fly ash is used as a partial replacement ingredient for Portland cement. Specifically, the study emphasizes the chemical composition and crystalline structure of the reaction products and their role in the compressive strength of concrete.

A number of standard concrete tests were conducted to determine the compressive strength, setting time, slump, air content, and unit weight. Additionally, a simple test for measuring the heat of hydration of the pastes was developed. X-ray diffraction (XRD) and scanning electron microscopy (SEM) were employed to help interpret the results of these tests, and determine the chemical composition and reaction products of the concrete.

MATERIALS

The Portland cement used in this research was produced by the Martin Marietta Company in Tulsa, Oklahoma. It is classified under ASTM C150 as Type I Portland cement and its cement mineral characteristics are given in Table I. The Blaine fineness of the cement was 3673 sq cm/gm. The fly ash was from the Oklahoma Gas and Electric Company Thermo Power Plant near Red Rock, Oklahoma. It is classified under ASTM C618 as Class C fly ash and its chemical analysis is presented in Table II.

TABLE I		TABLE II		
Cement Minerals of Portland Cement		Chemical Characteristics of Fly Ash		
Cement Materials	% of Total Weight	Oxides	% of Total Weight	
C_3S	55.8	SiO_2	34.00	
C_2S	17.4	Al_2O_3	22.56	63.85
C_3A	9.1	Fe_2O_3	7.29	
C_4AF	8.1	SO_3	2.75	
		CaO	26.88	
		MgO	5.06	
		Na_2O	1.46	
		Moisture Content	0.07	
		Loss on Ignition	0.45	

TESTING METHODOLOGY

General

To study the effect of fly ash on Portland cement mixtures, concretes and pastes were employed. The concretes were cast into 6 in x 12 in cylinders to measure compressive strength. Fragments of these cylinders were used for the SEM observations. The pastes were used primarily for the XRD and heat of hydration tests because the aggregates in concretes tended to interfere with these tests.

Mix Design

The mix design for the plain PC concrete was based on the Oklahoma Department of Transportation specifications for Class A concrete. The fly ash concrete mix designs were similar to the plain PC concrete mix design except that Portland cement was replaced by fly ash in amounts of 20, 30, 40, 50 and 100 percent by weight. The paste samples were proportioned in a similar manner to the concrete samples except that the water to solids ratio of the pastes was a constant 0.35. The compressive strength of the concrete was measured at 1, 3, 7, 14, 28, 90, 180, and 365 days.

Test Procedures

The compressive strength, time of set, mixing process, and fresh concrete properties tests were performed in accordance with standard ASTM procedures.

To measure the effect of the fly ash on the temperature rise of freshly mixed concrete, a simple test was devised. The test involved placing a temperature probe into freshly mixed paste and storing the paste in an insulated enclosure. A strip recorder connected to the probe continuously recorded the temperature of the paste for a period of 48 hours. For consistency, the test was always performed with the same amount of material (2500 gm of cementitious material and 875 ml of water) at the same water/solids ratio (0.35) and near the same initial temperature (70 to 75 deg F).

Pastes were used instead of concretes when it was found that the aggregate in the concrete so diluted the hydrating materials in the mix that little or no temperature rise could be detected.

Immediately following each series of compressive strength tests, fragments were removed from each broken cylinder (or paste cube of like age) for use in the SEM and XRD studies. From these fragments, quarter inch specimens of concrete matrix were produced for the SEM observations and finely ground concrete and paste powders were produced for the XRD analysis.

The first step in processing both SEM and XRD samples involved placing the fragments in a laboratory jaw crusher and reducing the sample size to approximately a quarter inch or less. The XRD samples were further ground to a powder in a laboratory pulverizer. Each type of sample was then doused with acetone to arrest the hydration process. After drying, the samples were again doused with acetone and allowed to dry before being sieved and stored in air tight containers.

The SEM specimens were prepared for viewing by first selecting an individual quarter inch fragment, oven drying the fragment at 110 deg C for one hour to remove any excess moisture, and then cementing it to an aluminum stub using a colloidal silver metal paint. A gold coating approximately 100 angstroms thick was then applied to the particle surface using a sputter coater.

PRESENTATION AND DISCUSSION OF RESULTS

Concrete Properties

The Class C high lime fly ash used in this study has a pronounced effect on all concrete properties tested. In most cases, the fly ash concrete behaved in a predictable manner, following patterns similar to those of other Class C fly ash concretes previously studied [8]. Table III presents typical concrete data obtained from the testing program.

As expected, the fly ash tended to retard the set of the concretes. The 100 percent fly ash mix, however, stiffened so rapidly it could not be discharged from the mixer and thus was not studied further. Interestingly, analysis of the heat of hydration data for test pastes revealed a similar pattern. The temperature of the 100 percent fly ash paste peaked at 106 deg F after only 30 minutes. This is compared to the plain PC paste which peaked at 118 deg F after 12 hours. Other fly ash-cement pastes displayed lower peak temperatures at later times as shown in Figure 1. A closer inspection of the data reveals that many of the fly ash-cement pastes actually displayed two peaks: the first presumably due to fly ash hydration and the second due to cement hydration. The relative heights of these peaks appear to be roughly proportional to the amounts of fly ash and cement present in the paste.

In a manner similar to the setting times, the early compressive strengths of the fly ash concretes were also retarded. Generally, in the early stages of hydration the fly ash tended to lower the compressive strength of the concretes in direct proportion to the amount of fly ash present in the concrete. By seven days, however, the average strength of the fly ash concretes always exceeded that of the plain PC mix. Furthermore, the concretes made with higher replacement fractions of fly ash tended to have greater compressive strengths than the concretes made with lower fractions of fly ash, Figure 2.

TABLE III

Test Concrete Data

	Plain	20% FA	30% FA	40% FA	50% FA
Quantities per cu. yd.					
Cement (lb)	564	452	395	338	282
Fly ash (lb)	0	112	169	226	282
Water (lb)	238	237	219	221	196
Fine Aggregate (lb)	1192	1192	1192	1192	1192
Coarse Aggregate (lb)	1187	1187	1187	1187	1187
Air Ent. Agent (ml)	243	243	264	290	299
W/C Ratio	.42	.53	.56	.66	.70
W/(C+F) Ratio	.42	.42	.39	.39	.35
Fresh Properties					
Slump (in)	2.50	2.25	2.25	2.00	2.00
Air (%)	5.2	5.0	5.2	5.3	5.2
Unit Weight (lb/cu ft)	144	145	145	143	144
Temperature (deg F)	64	69	68	70	64
Initial Set Time (hr:min)	5:36	9:24	12:06	17:30	18:12
Final Set Time (hr:min)	7:24	11:42	14:54	20:36	22:18
Strength* (psi)					
1 day	2098	2141	1661	1091	712
3 days	3759	3482	3261	3369	2847
7 days	4385	4462	4374	4840	4466
14 days	4500	5315	4943	5609	4921
28 days	4968	5243	5443	6395	6228
90 days	5362	4610	5553	6469	6333
180 days	5802	6676	6161	5911	6901
365 days	6349	7158	7131	8416	7590

Note: Setting time data are based on representative mortar mixes.

* Each number represents the average of three samples.

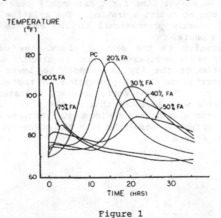

Figure 1

Heat of Hydration for Test Pastes

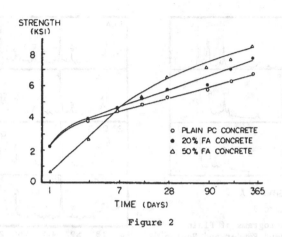

Figure 2

Compressive Strengths for 0%, 20%, and 50% Fly Ash Concrete

X-ray Diffraction

The diffractograms of paste samples were more informative than those of concrete samples since the pastes were not diluted by the aggregates in the concrete. For brevity, therefore, only the paste diffractograms will be discussed. Table IV lists the minerals and their respective symbols used in the diffractograms.

The plain PC paste diffractograms indicate the presence of a considerable amount of unhydrated cement minerals (C_3S, C_2S, C_3A, C_4AF) and three primary reaction products: calcium hydroxide, calcium carbonate, and ettringite. When the relative intensities of these peaks, Figure 3,

TABLE IV

Diffractogram Symbols

SYMBOL	MINERAL NAME
PC	PORTLAND CEMENT
P	PORTLANDITE
CC	CALCITE
E	ETTRINGITE
MS	MONOSULFOALUMINATE
Q	QUARTZ
G	GYPSUM
A	ANHYDRITE
L	LIME
C_3A	TRICALCIUM ALUMINATE

are compared over time, several trends can be noted. Two trends, well documented in previous research, are to be noted; first, the decline of the unhydrated cement minerals, second, the increase in the calcium hydroxide peaks. The ettringite peaks appear to be relatively constant as do the calcium carbonate peaks. The diffractograms probably overstate the amount of calcium carbonate because of the method of processing the samples.

The 20 percent fly ash pastes (Figure 4) display some similar trends to plain PC pastes. The amount of cement minerals tends to decrease over time and the amount of calcium hydroxide tends to increase slightly.

However, not all of the hydration processes in the 20 percent fly ash pastes appear to be similar to the plain PC pastes. For example, the amount of ettringite present in the plain PC pastes appears to be relatively constant, but in the 20 percent fly ash pastes the amount of ettringite appears to decrease. Furthermore, the intensity of the ettringite peaks in the 20 percent fly ash pastes appears to be greater than those of the plain PC pastes. While the amount of ettringite decreases over time, the amount of monosulfoaluminate increases. As expected, the ettringite in these pastes is converted into monosulfoaluminate.

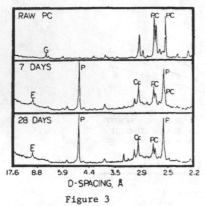

Figure 3

X-Ray Diffractograms of Plain
Portland Cement Powder and Pastes

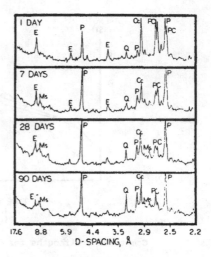

Figure 4

X-Ray Diffractograms of 20% Fly
Ash Pastes

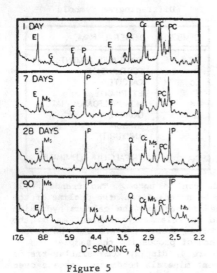

Figure 5

X-Ray Diffractograms of 50% Fly
Ash Pastes

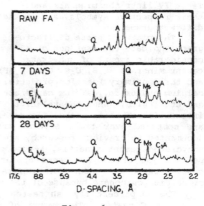

Figure 6

X-Ray Diffractograms of 100% Fly
Ash Powder and Pastes

The hydration processes of the 50 percent fly ash pastes (Figure 5) appear to be very similar to those of the 20 percent fly ash pastes. The unhydrated cement minerals slowly decrease over time as they did for the other pastes. However, the influence of the additional fly ash can be observed in the 50 percent fly ash pastes when comparing the relative intensities of C_3S and C_2S peaks with the C_3A peaks. The relative amount of the C_3A present in the early stages of hydration appears to be much greater in the 50 percent fly ash paste when compared to the other pastes. This increase in the relative C_3A content can be attributed to the fly ash, which contains a considerable amount of C_3A.

The major constituent of the 50 percent fly ash pastes at seven days and beyond is calcium hydroxide. The amount of calcium hydroxide appears to be relatively constant over this period.

Another hydration process similar to that of the 20 percent fly ash pastes is the conversion of ettringite to monosulfoaluminate. In the 50 percent fly ash pastes, however, this process is much more pronounced. At the early stages of hydration (one day) the intensity of the ettringite peak is much higher for the 50 percent fly ash paste than for the 20 percent fly ash paste. The conversion of the ettringite to monosulfoaluminate beyond this point is greatly accelerated when compared to the 20 percent fly ash paste. In fact, the ettringite peak is virtually nonexistent at 90 days.

Other minerals such as gypsum, periclase, and anhydrite are possibly present in the 50 percent fly ash paste as a result of the higher fly ash content. A large amount of quartz is believed to be present in the 50 percent fly ash paste as well. Positive identification, however, is somewhat clouded by both overlapping peaks and missing peaks. The other pastes also exhibited similar peaks of lesser intensity.

The diffractograms of the 100 percent fly ash pastes examined bear little resemblance to the other diffractograms. Instead, they appear to be similar only to the original fly ash powder diffractogram with peaks of the more soluble or reactive minerals such as C_3A, anhydrite, and CaO reduced by hydration and the new peaks from the reaction products such as ettringite, monosulfoaluminate, and calcite emerging from the background. In all of the 100 percent fly ash diffractograms, Figure 6, the quartz peak is dominant. The intensity of the quartz peaks, as well as the periclase peaks, appear to be relatively constant, suggesting that these materials do not participate in the hydration process. The cement mineral, C_3A, does appear to be reactive as expected. Even more reactive are the minerals anhydrite and CaO which are nearly undetectable after seven days of hydration.

SEM Observations

After examining the plain PC concrete specimens and the fly ash concrete specimens, it became apparent that the morphologies of the different fly ash concretes were similar. Thus, the fly ash concrete micrographs will be discussed as a single group.

In the one and seven day plain PC concretes, Figure 7, the most prevalent crystalline forms are foil-like structures tentatively identified as C-S-H gel or C_3A hydrate. These foil-like structures are similar in appearance to the hydration products of C_3A photographed by Plowman et al. [2] and the C-S-H gel photographed by Jawed et al. [3]. Frequently found with these foil-like structures are needle-like crystals tentatively identified as ettringite.

The 28 day micrographs of the plain PC concretes reveal fewer foil-like and needle-like structures. The needle-like structures observed at this period, Figure 8, tend to be more slender than the early needle-like crystals.

The 90 day specimens, Figure 9, also contain these slender needles

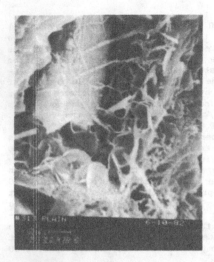

Figure 7. Plain P.C. Concrete
          @ 7-days (9000X)

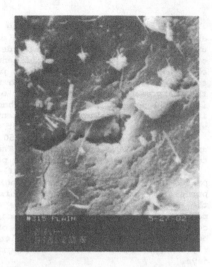

Figure 8. Plain P.C. Concrete
          @ 28-days (3000X)

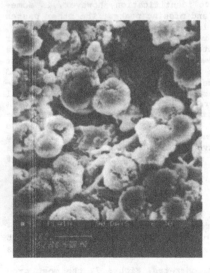

Figure 9. Plain P.C. Concrete
          @ 90-days (1500X)

Figure 10. 50% Fly Ash Concrete
           @ 1-day (3000X)

along with an abundance of an amorphous precipitate which appears to encrust a large portion of the specimen surface.

The one day specimens of fly ash concrete tend to display fewer crystalline forms compared to their plain PC counterparts. Among the few crystalline forms observed, are foil-like structures similar to those found in plain PC concretes. However, these structures in the fly ash concretes are much more difficult to detect. Equally difficult to detect are individual fly ash particles. As seen in Figure 10, nearly all of the fly ash particles at this stage of hydration are coated with a "duplex film" similar to that reported by Diamond [4] and Ghose [5]. This film tends to camouflage the fly ash spheres thus making identification difficult.

The 7 and 28 day specimens of fly ash concrete reveal a very different situation. Most of the fly ash particles are no longer totally encrusted with a duplex film, but are instead fairly smooth with no indication of a coating. These "clean" fly ash particles were apparently detached from the surrounding matrix when the sample was fractured. Several smooth voids, like the one in Figure 11, can be found where the fly ash had been "pulled out" of the surface. Such "pull out" features, as termed by Grutzeck et al. [6], were observed in samples seven days and older.

The 28 day micrographs also revealed the first signs of fly ash particle hydration. Many of the particles have an etched appearance, Figure 12, while others have an encrusted appearance, Figure 13.

The particle in Figure 12 typifies the etched morphology seen in both the 28 and 90 day fly ash concrete samples. Although this particle appears to be entwined in a web of interlocking crystalline fibers, it is possible that this reacted appearance is the result of the etching of the glassy component within the fly ash particle. Such etching would leave the less reactive crystalline material intact while removing the soluble glassy material.

The encrusted fly ash morphology in Figure 13 gives some indication of the depth of hydration surrounding the particle. In this micrograph, a portion of the reacted material of the particle has been broken away revealing a partial cross section of the particle, showing a distinct zone of hydration separating the unreacted portion of the fly ash particle from the surrounding matrix. Similar formations were reported by Grutzeck et al. [6] who observed a pattern of in situ hydration of the fly ash particle. According to Grutzeck's model, the glass sphere is gradually replaced by radiating bundles of C-S-H gel. Figure 14, a micrograph of a 90 day fly ash concrete sample, conforms to Grutzeck's model of fly ash particle hydration. In this case, the same "reacted zone" phenomenon is observed again; however, the fly ash particle is a partially broken cenosphere.

CONCLUSIONS

An explanation of the strength development of fly ash concrete may be divided into two categories: physical and chemical. Physically, the fly ash contributes strength by its size and spherical shape both of which help make the gel more compact and thus lower the water/solids ratio. Chemically, the development of strength appears to be linked to several factors.

One of the traditionally accepted explanations for Class F is that the pozzolanic reaction produces more C-S-H gel and thus enhances strength. Furthermore, it is generally accepted that the pozzolanic reaction of these Class F fly ashes reduces the level of calcium hydroxide. In this study, however, the Class C fly ash did not significantly reduce the level of calcium hydroxide. This does not imply that a pozzolanic reaction involving fly ash does not occur; rather it is indicative of the high levels of lime originally in the Class C fly ash. Thus it is presumed that a pozzolanic reaction involving fly ash occurs and contributes to strength.

Another chemical parameter that may be related to strength in fly ash concretes is the mineral ettringite. Several studies [2, 7] have concluded

82

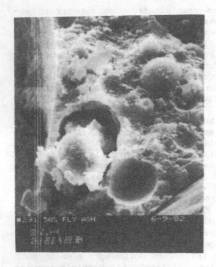

Figure 11. 50% Fly Ash Concrete
@ 28-days (3000X)

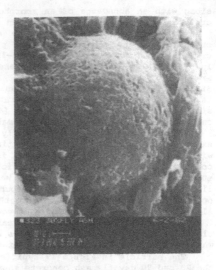

Figure 12. 30% Fly Ash Concrete
@ 28-days (7000X)

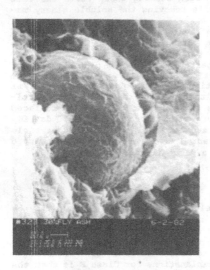

Figure 13. 30% Fly Ash Concrete
@ 28-days (7000X)

Figure 14. 50% Fly Ash Concrete
@ 90-days (3000X)

that ettringite inhibits setting in the early stages of hydration. Slower setting implies lower early strengths, as we observed. The X-ray diffractograms of the Class C fly ash pastes indicated high levels of ettringite in the one-day fly ash pastes. These high levels of ettringite are attributed to the presence of gypsum, anhydrite, lime, and C_3A in the Class C fly ash. Class F fly ashes typically do not have significant levels of the ettringite-forming minerals.

Ettringite may play a role in the later development of strength as well. The crystalline network formed during the setting stage is thought to provide the framework in which the C-S-H gel forms [7]. It is possible, then, that an early abundance of ettringite may provide more sites for the formation of the gel to commence and thus enhance strength. After the gel formation is well underway, however, the ettringite is considered as a source of weakness and its destruction is considered advantageous [7]. The X-ray diffractograms indicate that the level of ettringite in the fly ash pastes decreases with time and it is converted into the more stable monosulfoaluminate.

ACKNOWLEDGEMENTS

This work was supported by the Oklahoma Department of Transportation. The authors thank the Dolese Company, Oklahoma City, for the use of their laboratory, the Martin Marietta Company and the Oklahoma Gas and Electric Company for supplying the cement and fly ash, respectively. Also, thank are due Dr. T. Demirel of Iowa State University, Mr. W. Chissoe of Oklahoma University and Mr. C. Hayes of ODOT for their assistance during the conduct of this study.

REFERENCES

1. Baker, M.D., "Fly Ash Concrete: A Study of the Reaction Products Using X-ray Diffraction and SEM," M.S. Thesis, unpublished, University of Oklahoma, 1983.

2. Plowman, C., Cabrera, J.G., "The Influence of Pulverized Fuel Ash on the Hydration Products of Calcium Aluminates," Symposium Proceedings, Material Research Society, 1981.

3. Jawed, I., Skalny, J., "Hydration of Tricalcium Silicate in the Presence of Fly Ash," Symposium Proceedings, Material Research Society, 1981.

4. Diamond, S., "Characterization of Fly Ashes," Symposium Proceedings, Material Research Society, 1981, pp. 12-24.

5. Ghose, A., Pratt, P., "Studies of the Hydration reactions and Microstructure of Cement," Symposium Proceedings, Material Research Society, 1981, pp. 82-92.

6. Grutzeck, M., Roy, D., Scheetz, B., "Hydration Mechanisms of High Lime Fly Ash in Portland Cement Composite," Symposium Proceedings, Material Research Society, 1981, pp. 92-102.

7. Lea, F., "The Chemistry of Cement and Concrete," Chemical Publishing Company, Inc., Third Edition, 1970.

8. Berry, E., Malhotra, V., "Fly Ash for Use in Concrete - A Critical Review," ACI Journal, Proceedings V. 77, 1980.

AUTOCLAVE EXPANSION OF PORTLAND
CEMENT-FLY ASH PASTES

SCOTT SCHLORHOLTZ AND TURGUT DEMIREL
Iowa State University, Department of Civil Engineering, Ames, IA 50011

(Received 26 October, 1984; Accepted 13 February, 1985; Refereed)

ABSTRACT

The influence of the chemical composition of fly ash on the autoclave expansion of portland cement-fly ash pastes has been studied. The autoclave expansion tests were performed as described in ASTM C 151-81. Several of the cement-fly ash pastes tested exhibited severe expansive behavior. Free lime (CaO) was the only single variable that correlated well with the expansion observed in the autoclave tests. A multivariable model was generated in which lime, magnesium oxide, and tricalcium aluminate were the significant variables. The autoclave expansion of portland cement-fly ash pastes appears to be quite sensitive to the amount of free lime present in a given mixture.

INTRODUCTION

Volume stability of hardened portland cement concrete is a major requirement for the construction of long lasting concrete structures. Normally, one can distinguish between several mechanisms that lead to volume instability in hardened concrete such as changes in moisture content or temperature, mechanical forces, and chemical attack [1]. This paper deals only with the latter. The purpose of this paper is to show that the compound (or mineral) composition of fly ashes influence the volume stability (soundness) of portland cement-fly ash pastes that are subjected to autoclaving.

Unsoundness in portland cement is caused by the slow hydration of free magnesium oxide (MgO), free calcium oxide (CaO), and from ettringite formation from gypsum reacting with tricalcium aluminate or a calcium-aluminate-hydrate [2,3,4,5]. The latter, ettringite formation, is often considered as sulfate attack. The ettringite reaction will not be considered in this paper but the reaction must be mentioned because the autoclave test for measuring soundness is sensitive to one of the possible reactants, tricalcium aluminate ($3CaO.Al_2O_3$) [6]. Luckily, the autoclave test is much more sensitive to the presence of CaO and MgO; hence, it is not a bad approximation to ignore minor changes in the concentration of tricalcium aluminate.

THEORETICAL

The reactions influencing soundness are denoted as Equations 1 and 2.

$$CaO \ (s) + H_2O \ (l) \rightarrow Ca(OH)_2(s) \tag{1}$$

$$MgO \ (s) + H_2O \ (l) \rightarrow Mg(OH)_2(s) \tag{2}$$

When these reactions are slow but continue after the cement hardens, logic suggests that the soundness of a mixture containing these two compounds should be a function of the change in the partial molar volume ($\bar{V}$) of the products (solids) versus the reactants; therefore be dependent on the chemical potential ($\Sigma\mu_i$) of the reaction. Other factors such as particle size, crystallite size, reaction kinetics, and bonding forces undoubtedly may also influence soundness but we will ignore these other factors at the present time. For a pure component the partial molar volume, $\bar{V}$, is simply equivalent to the compounds molecular weight divided by the density of the compound. The

Table I. Partial molar volume calculations.

Compound	$\rho(g/cm^3)$	$\bar{V}(cm^3/mole)$	$\Delta\bar{V}(cm^3/mole)$ (solids)	% increase $\bar{V}$ (solids)	$\Delta G°(kcal/mole)$
CaO	3.32	16.89			-144.4
Ca(OH)$_2$	2.24	33.08	16.18	96	-214.33
MgO	3.58	11.26			-136.13
Mg(OH)$_2$	2.36	24.71	13.45	119	-199.27

change in partial molar volume of reactions 1 and 2 is calculated from data tabulated in various sources [7,8]. Table I summarizes pertinent data and the results of the partial molar volume calculations. The percent increase in partial molar volume due to hydration is also listed in Table I, along with free energies of formation ($\Delta G°$ at standard state). The partial molar volume of water consumed in the reaction is excluded in calculation of the increase in partial molar volume. The reason for this omission is the fact that the water of hydration reaches the lime or periclase encapsulated in the hardened cement matrix by diffusion from the surroundings. Hence the total volume change causing the destructive expansion would be due to partial molar volume changes listed in Table I.

The differential Gibbs free energy (dG) is defined by the following equation as a function of temperature (T), pressure (P), and the number of moles (N_i) of each component present:

$$dG = -SdT + VdP + \sum_{i=1}^{K} \mu_i dN_i \qquad (3)$$

which is the fundamental equation of chemical thermodynamics [9], where S is entropy, V is volume, K is the number of components present, and μ is the chemical potential. We can further simplify Equation 3 by allowing only isothermal processes in our system. Thus, Equation 3 reduces to

$$dG = VdP + \mu_i dN_i + \mu_k dN_k \qquad (4)$$

for a two component system.

Now let us construct an ideal (infinitely rigid) system such as the one shown in Fig. 1a as an analog to a cement paste matrix surrounding a crystal of CaO (Fig. 1b). Assume that the system as a whole is so much larger than the sample (CaO crystal) that we can neglect the temperature change due to the hydration of CaO. Since G is a state function (independent of path) we can construct a series of simple paths which describe the changes in the system, sum them up, and solve for the pressure that must be exerted by the piston in Fig. 1a to hold the system in equilibrium (i.e., at constant volume).

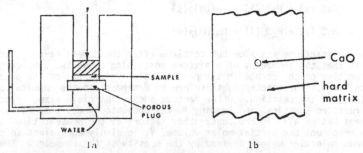

Fig. 1. An analog for the hydration of CaO in a hardened cement paste.

Table II. Reaction equilibrium calculations.

Step	Reaction			Change in Gibbs free energy (per mole)
1	$CaO(s)$ (P_o,T_o)	$+ H_2O(l)$ (P_o,T_o)	$\rightarrow Ca(OH)_2(s)$; (P_o,T_o)	$\Delta G_1 = \Delta G°$
2	$CaO(s)$ (P,T_o)	$\rightarrow$	$CaO(s)$; (P_o,T_o)	$\Delta G_2 = \int_P^{P_o} \bar{V}_{CaO}dP$
3	$Ca(OH)_2(s)$ (P_o,T_o)	$\rightarrow$	$Ca(OH)_2(s)$; (P,T_o)	$\Delta G_3 = \int_{P_o}^{P} V_{Ca(OH)_2}dP$
4	$CaO(s)$ (P,T_o)	$+ H_2O(l)$ (P_o,T_o)	$\rightleftarrows Ca(OH)_2(s)$; (P,T_o)	$\Delta G_4 = 0$

Table II summarizes the steps involved. In step 1, we allowed the reaction to take place at standard temperature (T_o = 298°K) and pressure (P_o = 1 atmosphere). Steps 2 and 3 deal with the isothermal decompression and compression of CaO and $Ca(OH)_2$ respectively. The integrals can be simplified by assuming that the partial molar volume is independent of pressure which is the usual assumption made for solids. Step 4 describes the reaction at equilibrium (i.e. differential adjustment of P determines the direction of reaction); therefore $\Delta G_4 = 0$. The sum of steps 1 through 3 is equivalent to step 4 or $\Delta G_4 = \Delta G_1 + \Delta G_2 + \Delta G_3$. Substituting in the various terms from Tables I and II, and solving for ΔP yields

$$\Delta P = \frac{-\Delta G°}{\bar{V}_{Ca(OH)_2} - \bar{V}_{CaO}} \tag{5}$$

A similar expression can be derived for MgO. Substituting in the numerical values yields a ΔP of 5 x 10^5 (3.5 x 10^6 kPa) psi for calcium oxide and a ΔP of 3 x 10^5 (2 x 10^6 kPa) psi for magnesium oxide. Obviously no material can withstand pressures of this magnitude (with the exception of our ideal apparatus) so in general strain must take place to reduce the stress to a reasonable value. The strain manifests itself as unsoundness. Because of the idealizing assumptions made in its derivation Equation 5 can be applied only in a semiquantitative sense to a real cement matrix to estimate the order of expansion pressure.

Typically, since soundness problems do not become evident for months and possibly years because of slow hydration rates, industry uses accelerated methods of hydration to shorten the time required to identify expansion prone cements. The autoclave method described in ASTM C 151 is one of these accelerated methods [10]. The increase in temperature and pressure in the autoclave generally causes the kinetic problems normally associated with the hydration of hard-burnt (an industrial term describing lack of reactivity) lime and periclase to become negligible letting the thermodynamics of the system, which strongly favors the hydration reactions, be the controlling factor. The autoclave test has been used in this investigation as a means for the rapid determination of the relationship between the concentration of expansive components in cement pastes and their tendency to expand. The relevance of limits set by ASTM to actual field performance are not taken into consideration in this investigation. Other variables such as crystal and particle sizes of the expansive components which undoubtedly play an important role in expansivity of hardened cement matrix are not included in the present phase of the study.

EXPERIMENTAL PROCEDURE AND RESULTS

The experimental results presented in this paper were not produced by a single experiment, rather they are taken from results generated in the last two years by the Materials Analysis and Research Lab at Iowa State University which is actively engaged in monitoring the quality of all Iowa fly ashes. We are going to emphasize a few fly ashes that exhibit severe expansive behavior, keeping in mind that these fly ashes are uncommon compared to the majority of Iowa fly ashes that exhibit excellent soundness. The presentation of experimental results will be divided into two different groups. The first consists of an initial, in-depth study of the autoclave expansion characteristics of three fly ashes. The second group simply adds more evidence to support the theory of soundness developed earlier in this paper.

Component and elemental analysis were performed on a Siemens D-500 diffractometer and a LC-200 sequential spectrometer, respectively. Both systems were coupled to a PDP 11/03 mini computer for data accumulation and processing. Manual operation of the D-500 diffractometer was employed when needed. The details of the calibration, operation, and sample preparation are described in reference [11]. Elemental analysis of the three portland cements is summarized in Table III; the results of the determination of free lime, periclase, and autoclave expansion are also included. Table IV summarizes the elemental and compound analyses of the fly ashes.

Quantitative x-ray diffraction (QXRD) was used for the determination of lime and periclase for both the fly ashes and the portland cements. The method of Known Addition was used for QXRD with sodium chloride as the internal standard. Details of this method are described in earlier publications [12,13]. Bogue's equations were used to determine the tricalcium aluminate (C_3A) content of the portland cements; QXRD was used to measure the C_3A content of the fly ashes.

Group I

In this investigation, three common Iowa fly ashes (Table IV, columns 1, 2 and 3) and one Type I portland cement (Table III, column 1) were used. Substitution of fly ash for portland cement ranged from 20% (as required in ASTM C 311) to 50% by weight. Fifty-two specimens were molded and subjected to autoclaving. Included in the 52 samples were eight control samples composed only of portland cement. All autoclave samples were made, cured, and tested in accordance with ASTM C 151 [10]. Distilled water was used throughout the experiment.

All of the expansion data for the cement-fly ash mixtures were divided by the average expansion observed for the control specimens, this normalization will be called the relative expansion (RE). The Statistical Analysis System (SAS) was used for investigating possible relationships between the relative expansion and the chemical composition of the cement-fly ash mixtures [14]. Linear regression analysis was applied to all variables that appeared to affect the relative expansion, the functional relationship was expressed qualitatively by plotting all the variables versus relative expansion. Regression analysis indicates that free calcium oxide has by far the most significant influence on the relative expansion of the cement-fly ash pastes. In fact, it was the only single variable that was a true function of relative expansion. Figure 2 shows a plot of relative expansion versus % free lime. The data shown in Fig. 2 was analyzed by two different methods. First, the whole plot was fitted with a fourth order polynomial. The polynomial provided a reasonable fit (coefficient of determination, R^2, equal to 0.93) for a portion of the data but such a model was not adequate to predict relative expansions for low concentrations of lime. Thus, the second method of analysis

Table III. Chemical characteristics of the three Type I portland cements used in this study.

Oxide	1 IDOT blend	2 Lehigh	3 Monarch
CaO	63.7	n/a	63.0
Al_2O_3	4.3	n/a	4.0
Fe_2O_3	2.4	n/a	2.9
SiO_2	22.1	n/a	20.3
SO_3	2.6	n/a	2.5
K_2O	0.6	n/a	0.4
MgO	3.6	n/a	2.3
Na_2O	0.2	n/a	0.2
P_2O_5	0.1	n/a	0.1
Free CaO: QXRD	0.2	1.0	0
Free MgO: QXRD	2.0	1.5	0
C_3A[a]	7.8	n/a	5.5
Autoclave expansion	0.06	0.18	-0.02

[a]Calculated from Bogue's equations

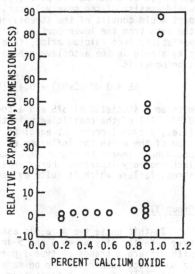

Fig. 2. Results obtained from autoclaving Group I specimens.

Table IV. Chemical and mineral characteristics of the five fly ashes used in this study.

a) Elemental analysis (QXRD)

Oxide	Fly Ash 1 Neal 3	2 Neal 4	3 Omaha North	4 Neal 2	5 Sherburn
SiO_2	50.5	32.1	50.0	52.5	44.3
Al_2O_3	18.1	19.7	24.1	18.1	18.0
Fe_2O_3	7.87	5.79	18.2	6.22	5.55
CaO	13.6	29.5	1.98	17.3	20.1
MgO	3.14	7.74	1.31	3.20	3.97
TiO_2	0.66	1.17	0.65	0.62	0.72
Na_2O	0.51	2.23	0.56	0.33	3.30
K_2O	1.24	0.30	1.06	1.72	0.99
SO_3	---	---	---	1.61	2.65

b) Compound analysis (QXRD)

Compound	Neal 3	Neal 4	Omaha North	Neal 2	Sherburn
CaO	2.3	1.1	0.2	5.6	2.5
MgO	1.0	3.0	0	0	1.0
$3CaO.Al_2O_3$	0	3.9	0	trace	0.5

consisted of breaking the plot (Fig. 2) into two distinct parts. One part would consist of the nearly linear, lower portion of the plot while the other part would consist of the steeply inclined portion of the plot. Regression of the data from the lower portion of the plot of Fig. 2 indicates that three variables, free calcium oxide, free magnesium oxide, and tricalcium aluminate, play a role in the autoclave expansion of the different mixtures. The regression model is

$$RE = 3.00 (CaO\%) + 0.43 (MgO\%) + 0.26 (C_3A\%) - 2.52 \qquad (6)$$

with an F statistic of 375 and a coefficient of determination, R^2, equal to 0.97. All of the coefficients in this model are significant at the 5% level (i.e., a type I error, α, equal to 5%). Even on the well behaved, lower portion of the curve the influence of free lime on the relative expansion is quite pronounced (note the magnitude of coefficients in Equation 6). Equation 6 does lack one very important characteristic in that it can not predict the point of abrupt failure which is quite evident in Figure 2.

Group II

In this phase two more fly ashes were identified which exhibited expansive behavior (fly ashes 4 and 5 in Table IV). Unfortunately they were tested with different brands of type I portland cement (cements 2 and 3 in Table III) so no direct comparison can be made between any of the expansive fly ashes used in both phases. Substitution of fly ash for portland cement ranged from 5 to 20% for fly ash number 4 and from 13 to 40% for fly ash number 5. A total of 28 specimens were molded and subjected to autoclaving as described earlier in this paper.

Figure 3 shows a plot of auotclave expansion versus percent fly ash for the three expansive fly ashes. It should be noted that failure is always typically quite abrupt, an increase of 1% fly ash can cause a transition from passing the specification (<0.8% expansion) to failure. This abrupt expansive behavior is very similar to results reported in an earlier study on the expansion characteristics of portland cement [15]. Such sensitivity may be quite important to on-site batch plants which may not be able to maintain adequate quality control.

Of the three expansive fly ashes the Neal 2 fly ash is a rather special one. Not only is it the ash that failed to meet the ASTM specification for autoclave expansion at the specified (20%) replacement but it also contains no significant amounts of periclase or tricalcium aluminate. Figure 4 shows the X-ray diffraction pattern of the Neal 2 fly ash. Thus, for the Neal 2 ash, there is no doubt that free lime is the single cause of failure in the autoclave test.

If we again assume that autoclave expansion is only dependent on the amount of free lime present in a given mixture then we can plot all of the data (both Group I and Group II) on a single graph. Figure 5 is such a plot. At present, although the scatter of data in Figure 5 seems to be inadequate for fitting a single valued function to, the deviation from such a function appears to be within the present accuracy of measurements. It should be noted that the results depicted in Figure 5 include three different portland cements and five different fly ashes, a total of 80 data points. As we refine the measuring techniques that we are currently using we will be in a better position to decide which of the remaining variables (crystallite size, particle size, etc.) need to be included in our soundness model.

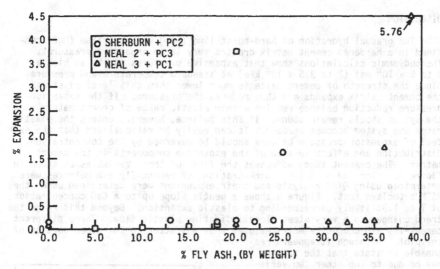

Fig. 3. Percent expansion versus percent fly ash for the
three fly ash-cement pastes that failed the autoclave test.

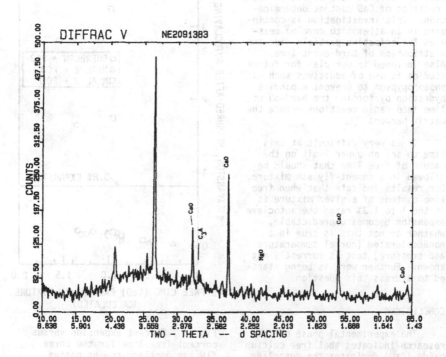

Fig. 4. X-ray diffraction trace of Neal 2 fly ash (air dry).

DISCUSSION

The gradual hydration of hard-burnt lime (CaO) and periclase (MgO) confined in a hardened cement matrix creates very large expansive pressures. Thermodynamic calculations show that expansive pressures can be as high as 3 to 5 x 10^5 psi (2 to 3.5 x 10^6 kPa) at standard temperature and pressure. Since the strength of cement paste is much lower than this level of stress the cement matrix expands and thus reduces the pressure. If the expansion-pressure reduction balance remains within elastic range of cement matrix the system should remain sound. If this balance, however, enters the plastic range the system becomes unsound. It can easily be rationalized that the extent of expansion-pressure balance should be governed by the concentration, distribution and effective size of the expansive components in the cement matrix. The present study addresses the first of these variables as shown in Figure 5. Lime and periclase concentrations of cement-fly ash matrices were determined using QXRD analysis and their expansions were determined using the ASTM autoclave test. Figure 5 shows a gentle slope up to a CaO concentration of 1 to 1.3% likely corresponding to elastic expansion. Beyond this range the trend changes to a very steep slope indicating plastic flow. Three different cement-fly ash composite matrices used show a scatter in the steep portion of the plot. Although it appears reasonable to state that the scatter may be due to the other two variables mentioned above, namely, distribution and effective size of encapsulated hard-burnt lime, the scatter observed is also within the precision of CaO content determination. This investigation is continuing in an attempt to develop measures for effective particle size and distribution of hard-burnt lime. Also included in our plan for future studies is use of additives such as phosphogypsum to prevent expansive hydration by forcing the hard-burnt lime into rapid reactions before the matrix hardens.

It is very difficult at this time to set an upper limit on the amount of free lime that should be allowed in a cement-fly ash mixture. Our results indicate that when free lime content of a given mixture is in the 1 to 1.3% range the autoclave expansion becomes unpredictable, whether or not this is true in a nonaccelerated (normal temperature and pressure) test is currently not known. Further work is being started to address this question.

CONCLUSIONS

The experimental phase of this research indicates that free calcium oxide (CaO) dominates the autoclave expansion of portland cement-fly ash pastes. This conclusion is also

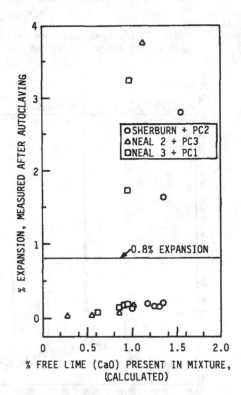

Fig. 5. Percent expansion versus percent free lime for the three fly ash portland-cement pastes that failed the autoclave test.

supported by a thermodynamical treatment of the hydration of CaO and MgO. There can be exceptions to this observation (i.e., some fly ashes high in CaO may exhibit low expansive properties) because many fly ashes, particularly Class C ashes, have other components present that may influence the reactivity of CaO. Tricalcium aluminate and MgO do influence the autoclave expansion of portland cement-fly ash pastes (as indicated by Equation 6) but to a much smaller magnitude than free lime.

REFERENCES

1. J. L. Sawyer, "Volume Change," ASTM STP 169B (1978).
2. P. K. Mehta, "History and Status of Performance Tests for Evaluation of Soundness of Cements," ASTM STP 663 (1978).
3. F. M. Lea, The Chemistry of Cement and Concrete, Chemical Publishing Co. (1971).
4. R. H. Bogue, The Chemistry of Portland Cement, Reinhold Pub. Co. (1955).
5. V. S. Ramachandran, R. F. Feldman, and J. J. Beaudoin, Concrete Science, Heyden Pub. Co. (1981).
6. R. N. Young, Proc. of ACI, 34 (Sept.-Oct., 1937).
7. R. Weast, editor, Handbook of Chemistry and Physics, CRC Press (1978).
8. L. Berry, editor, Selected Powder Diffraction Data for Minerals, JCPDS (1974).
9. I. N. Levine, Physical Chemistry, McGraw-Hill Pub. Co. (1978).
10. American Society for Testing and Materials, 1981 Annual Book of ASTM Standards, Part 13 (1981).
11. M. L. Mings, "Effect of Fly Ash Reactivity on Fly Ash Paste," Unpublished M.S. Thesis, Iowa State University (1982).
12. M. L. Mings, S. Schlorholtz, J. M. Pitt, and T. Demirel, Transportation Research Record, 5 (1983).
13. J. M. Pitt, M. L. Mings, and S. Schlorholtz, Transportation Research Record 941 (1983).
14. J. Helwig and K. Council, editors, SAS Users Guide, SAS Institute (1979).
15. H. F. Gonnerman, W. Lerch, and T. M. Whiteside, PCA Laboratory Bulletin 45 (June 1953).

EFFECTS OF FLY ASH AND SUPERPLASTICIZERS
ON THE RHEOLOGY OF CEMENT SLURRIES

ELIZABETH L. WHITE*, MARIA LENKEI**, DELLA M. ROY,* AND FERENC D. TAMÁS***
  *Materials Research Laboratory, The Pennsylvania State University,
   University Park, PA 16802;
 **Central Research and Design Institute for Silicate Industry, H-1034
   Budapest, Bécsi Út 126, Hungary;
***Department of Silicate Chemistry, Veszprém University, H-8201 Veszprém,
   P.O.B. 158, Hungary

(Received 29 November, 1984; accepted 8 February, 1985; Refereed)

ABSTRACT

    Cementitious slurries composed of an oil well cement, a high calcium fly
ash, a low calcium fly ash, and three commercially available superplasticizers
(two different sulfonated naphthalene formaldehyde condensates and a sulfonated melamine formaldehyde condensate)were mixed to contrast the two fly
ashes and to determine the effectiveness of each of the superplasticizing
agents. Most commercial superplasticizers and cements are relatively expansive;therefore a partial substitution by fly ash and other by-products represents a substantial savings in both quantity of chemical admixture required
and energy consumption for the manufacture of cement. In the cement/fly ash
mixtures of 100/0, 90/10, 60/40, 40/60, and 10/90, with both high calcium fly
ash and low calcium fly ash, the mixture containing the low calcium fly ash
was consistently less workable. The rheological properties of the high vs.
low calcium fly ash mixtures were controlled by the differences in fly ash
particle size and the presence of irregular large particles, rather than by
the differences in chemistry between the two. The low calcium fly ash was
the coarser material.

INTRODUCTION

    The rheological properties of oil well (Class H) cement slurries were
modified with both chemical admixtures of the superplasticizer family, salts
of sulfonated melamine formaldehyde condensate (SMFC) and sulfonated naphthalene formaldehyde condensate (SNFC) with both low calcium and high calcium
fly ash reactive agents. A preliminary study [1] was made on the rheological
properties of Type 1 and an oil well cement (Class C) slurries with two SNFC
superplasticizers (one with a retarder) and one SMFC superplasticizer. In
that study a high speed mixing procedure was used throughout the testing and
no fly ash was used.

    It is of great interest to use by-products, such as fly ash, because
they impart favorable rheological properties in combination with cement and
also because of their abundant supply as residue from coal-fired furnaces.
This study includes a low calcium fly ash (our code B25 at $CaO = 1.8\%$)
and a high calcium fly ash (our code B62 at $CaO = 27.8\%$) to span the more
probable fly ash compositions. The low calcium fly ash was derived from a
bituminous coal whereas the high calcium fly ash was derived from a subbituminous coal.

    The effectiveness of the chemical admixtures, sodium salts of sulfonated
melamine formaldehyde condensate (our code A73) and two sulfonated naphthalene
formaldehyde condensates, our code A41, manufactured in the United States,
and our code A74, manufactured in Japan, were compared in the presence of 100
to 10 percent cement and respectively 0 to 90 percent fly ash, using both the
low calcium and the high calcium fly ash. The purpose of this research was

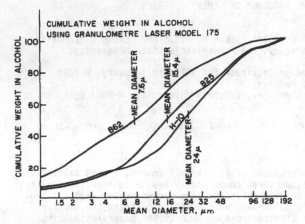

Figure 1. Cumulative weight-size distribution in alcohol of the Class H cement (H-10), the high calcium fly ash (B62), and the low calcium fly ash (B25).

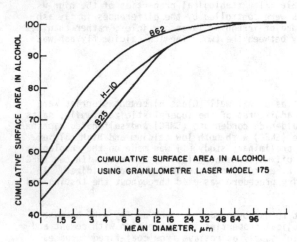

Figure 2. Cumulative surface area in alcohol (expressed as proportion of total) of the Class H cement (H-10), the high calcium fly ash (B62), and the low calcium fly ash (B25). B.E.T. for H-10 = 0.67; B25 = 0.85; B62 = 1.06. m²/g

to compare the rheological properties of these mixtures with respect to the type and quantity of chemical admixtures needed vs. the calcium content (or chemical composition) of the fly ash available.

MATERIALS AND METHODS

The weight and surface area distributions for the cement and each fly ash were measured using a C.I.L.A.S. Granulometre Laser device, model 175 (manufactured by Compagnie Industrielle des Lasers, Marcoussis, France). Before measurements were made, all particles larger than 200 μm were removed from the high calcium fly ash. As shown in Fig. 1 the mean diameter for the high calcium fly ash was 7.6 μm; the mean diameter for the low calcium fly ash was 15.4 μm; and the mean diameter for the Class H cement was 24 μm. The specific surface area of the high calcium fly ash was greater than the Class H cement or the low calcium fly ash, as shown in Fig. 2.

Scanning electron micrographs of the two types of fly ash are shown in Fig. 3. The matrix material present in the low calcium fly ash does not appear in the high calcium fly ash (compare Fig. 3a with 3b).

The chemical composition for each fly ash and the Class H cement are given in Table I, showing that the CaO content for the low calcium, the high calcium, and the Class H cement increase from 1.82 to 62.99. The chemistry of fly ash is discussed in reference [2].

EXPERIMENTAL PROCEDURES

The dry cement, fly ash, and dry superplasticizers were preblended; the required volume of deionized water was added to the dry mixture. The time when the liquid and dry components were mixed was defined as time = 0.0. Forty milliliters of slurry were tested using a Haake coaxial cylindrical Rotovisco Model RV 3 viscometer; 10 mL of slurry were tested using the Rheotron coaxial cylindrical Model 8021-01. Water/solid ratio was fixed at 0.33.

For the Haake measurements, the slurries were hand mixed for 30 seconds; testing began after 1.00 to 1.50 minutes depending on the fluidity and thus the ease in transferring the sample to the measuring cup. The viscometer was cycled from 0 to 40 sec^{-1} for 4 cycles with start times of 1.00, 3.50, 5.75, and 7.50 minutes. The constants used for these measurements were the following:

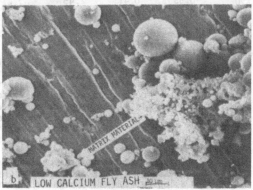

SHEAR FACTOR (M) = 0.78
STRESS FACTOR (A) = 3.76
MEASURING SYSTEM = MK50
    Measuring Head
ROTOR AND CUP = SVIIP -
    serrated cup and rotor
MAXIMUM TORQUE = 0.490 N-cm
    (50 cm-g)
TORQUE RANGE = 3500 Pa

Figure 3. Scanning electron photomicrographs of (a) the high calcium and (b) the low calcium fly ash, showing larger particle size and irregular fragments in (b).

For the Rheotron measurements, the slurries were hand mixed for 2.5 minutes; testing began after 3.5 minutes had elapsed. The viscometer was cycled from 0 to 50 sec^{-1} for 3 cycles. Each cycle lasted for 40 seconds, much faster than the Haake tests. The constants used for the Rheotron measurements were the following:

SPRING FACTOR = A   MEASURING SYSTEM = C5/C
SHEAR FACTOR = 0.897   MEASURING RANGE = 10
STRESS FACTOR = 4.376   GEAR = 10
NUMBER STEPS = 240/40 seconds
Apparent Viscosity Range: η = 5 to 4×10^8 mPa-sec
Shear Stress Range: τ = 2.2 to 2.2×10^4 Pa
Shear Rate: D = 0.05 to 1×10^3 sec^{-1}

A computer driven program controlled the measurements and then calculated the flow properties for Newton/Bingham and Casson behaviors. The program, written for the Rheotron by Brabender (4100 Duisberg, West Germany), used the following relationships: Newtonian flow: $\eta = \tau/D$; Bingham flow:

Table I. Chemical composition of Class H cement, high calcium fly ash, low calcium fly ash, and each of the compositions tested in this study.

	H-10 cement	B62 HFA*	B25 HFA**
SiO_2	22.24	36.8	50.2
Al_2O_3	3.58	18.10	27.00
TiO_2	0.10	1.55	1.40
Fe_2O_3	3.94	6.21	13.80
MgO	3.87	4.47	0.84
CaO	62.99	27.83	1.82
MnO	0.005	0.039	0.034
Na_2O	0.13	1.22	0.24
K_2O	0.53	0.41	2.45
LOI, 1000°C	0.66	0.55	--

*HFA: High Calcium Fly Ash (B62).
**LFA: Low Calcium Fly Ash (B25).

$n = (\tau-\tau_0)/D$, where τ_0 = yield stress.

The rotor and cup were serrated for the Haake instrument measurements; they were smooth-walled for the Rheotron viscometer measurements. In a few experiments, slippage was a problem using the smooth walls for cementitious materials, however, slippage did not occur usually until the third cycle using the Rheotron viscometer.

The testing period using the Haake spanned those measurements taken with the Rheotron. The flow behavior on the Rheotron took place in a time equivalent to the second cycle on the Haake.

RESULTS

Pumpability and workability characteristics of cementitious slurries and mortars are a function of the rheological properties of the mixture. Workability of concrete mixtures refers generally to the amount of stress necessary to "work" (mix) a particular mixture, while pumpability of a grout or concrete also includes the concept of yield stress, i.e., the more pumpable the lower the stress to get the mixture moving; i.e. (zero or) low-yield stress materials are flowing or pumpable mixtures. A comprehensive study on the effects of various admixtures of superplasticizers, by-products, and aggregates has been published by White and Roy [3]. Cementitious mixtures containing industrial by-products, Roy et al. [5], have been shown to have the most favorable physical properties [4,5]. For this reason a low calcium and a high calcium fly ash by-product were each mixed in various proportions with a Class H cement and three chemical admixtures of the superplasticizer family. Changes in the shear stress-shear rate relationships, the yield stress, and the plastic viscosity were related to the fly ash admixture type and concentration as well as the superplasticizer type and concentration.

Haake Rotovisco RV3 Results

The rheological properties of the 27 mixtures measured are summarized in Figs. 4-7. For those mixtures containing high calcium fly ash (B62) the quantity of superplasticizer was fixed at 0.05g/10ml liquid/40g sample whereas the quantity of superplasticizer for the mixtures containing low calcium fly ash (B25) was varied from 0.10 to 0.30g/10ml liquid/40g sample.

Two to six times as much superplasticizer was required for the low calcium fly ash mixtures compared with the high calcium mixtures in order to obtain similar rheological flow properties. For 40 percent fly ash, the quantity of superplasticizer had to be increased four-fold to make the low calcium (at CaO = 1.8%) and high calcium (at CaO = 27.8%) mixtures have similar viscosity. At the 60% fly ash mixture a six-fold increase in superplasticizer was required for the low calcium mixture to produce rheological properties similar to those for the high calcium mixtures. At 90% fly ash the viscosities of the high calcium mixtures (at 0.05g/40g sample) were higher than those of the low calcium fly ash mixtures (at 0.30g/40g sample).

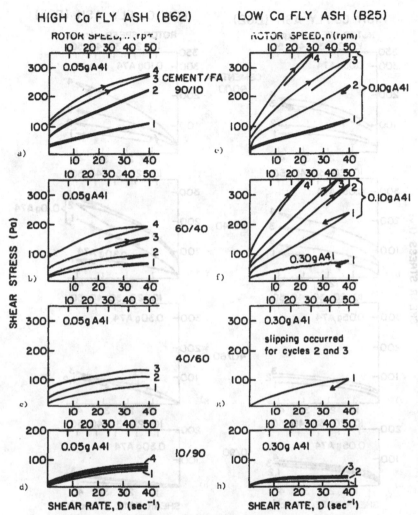

Figure 4. Rheological properties (measured on the Haake viscometer) of Class H cement (H-10) with different proportion of high calcium and low calcium fly ash mixed with SNFC (A41) superplasticizer.

Another interesting observation was that an increase in the proportion of either type of fly ash beyond 40% brought about a decrease in the viscosity of the mixture, so that the mixtures with more fly ash and less cement were more pumpable. Thus the water/salts ratio could be reduced for a comparable increase in the fly ash content irrespective of the type of superplasticizer or quantity of superplasticizer used. This was not self-evident at the start of this study.

For the high calcium mixtures it was apparent that the yield stress and the hysteresis (structural changes within the mixtures) loops were also independent of the superplasticizer type (Figs. 4-6, a-d). On the other hand, a

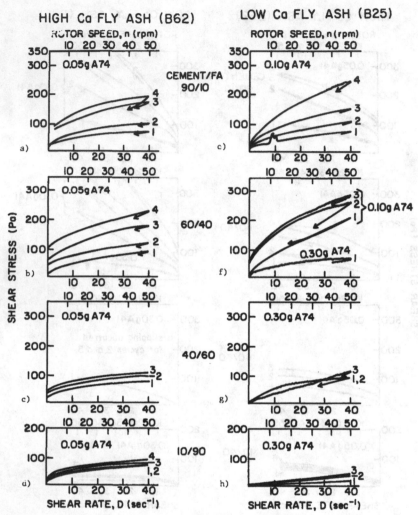

Figure 5. Rheological properties of Class H cement (H-10) with different proportion of high calcium and low calcium fly ash mixed with SNFC (A74) superplasticizer.

comparison between types of superplasticizers for the low calcium fly ash mixtures showed extensive variability. The SMFC-containing mixtures (Fig. 6e-h) had very irregular flow behavior, while maintaining thixotropic flow properties throughout. The SNFC mixtures were more pumpable than the melamine counterparts at a constant water/cement and superplasticizer content.

As shown in Fig. 7 the addition of 0.05g vs. 0.30g of superplasticizer admixture had very little impact on the viscosity of the non-fly ash mixtures at the same w/c as done with the fly ash mixtures. Again, as with the fly ash-containing mixtures, the SNFC superplasticizers were more effective than the SMFC superplasticizers in producing a pumpable mixture.

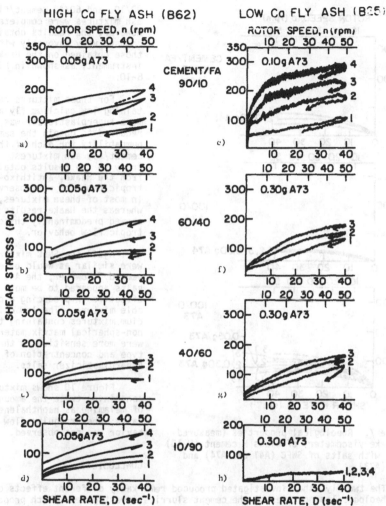

Figure 6. Rheological properties of Class H cement (H-10) with different properties of high calcium and low calcium fly ash mixed with salts of SMFC (A73) superplasticizer.

Rheotron 8021 Results

The viscosity in Pa-sec for the 15 mixtures tested on the Rheotron 8021 (in Figs. 8-10) is shown in the high and low calcium fly ash mixtures with the same three superplasticizers. Whereas the Haake data show the shear stress vs. the shear rate relationships, these data show viscosity vs. shear rate. For these data the quantity of superplasticizer was kept constant for each type of super-plasticizer with the SNFC quantity fixed at 0.05g/10ml of liquid/30g of cement + high calcium fly ash or low calcium fly ash. For the SMFC superplasticizer salts, the quantity was kept at 0.10g/10ml liquid/30g (cement + high calcium) fly ash or low calcium fly ash. For this set of measurements only the 100/0,

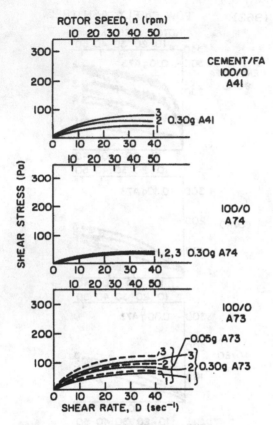

Figure 7. Rheological properties (measured on Haake viscometer) of Class H cement (H-10) mixed with salts of SNFC (A41 and A74) and SMFC (A73).

90/10, and 60/40 cement/fly ash mixtures were completed. However, the results obtained compare very favorably with those obtained on the Haake instrument, as shown in Figs. 8-10.

For those mixtures containing low calcium fly ash more superplasticizer was required to obtain the same pumpability for each of the cement/fly ash mixtures. Contrary to the results obtained from the Haake, antithixotropic behavior was observed in most of these mixtures, whereas the Haake results showed predominantly thixotropic flow behavior.

Again the SNFC mixtures were similar as would be expected. However, the A41 salts appeared to be more effective in producing a pumpable mixture. The low calcium mixtures containing the non-spherical matrix material were more sensitive to the type and concentration of the superplasticizer salts.

Figure 10 shows mixtures containing twice the amount of melamine as naphthalene salts; the irregular flow behavior was not observed.

CONCLUSIONS

The two fly ashes investigated produced remarkably different effects on the rheological properties of the cement slurries investigated. With proportions of liquid to solid kept constant, the substitution of the fine (and regular) particulate high-calcium fly ash for part of the coarse oil well cement generated lower viscosities/total shear stress values, while the substitution of the irregular particulate coarser low-calcium fly ash had the opposite effect; unless high amounts of dispersing superplasticizers were also added. The quantity of superplasticizer required to develop equivalent viscosity in mixtures of a coarse Class H oil well cement mixed with a high and low calcium fly ash correlated with the particle size distribution. Two to six times as much superplasticizer was required for the coarse low calcium fly ash mixtures than for comparable high calcium mixtures to obtain the same viscosity. The results of apparent viscosity measurements made by two different instruments in two different laboratories are in general agreement, giving greater credibility to the data.

With other factors kept constant, the chemical composition of the fly ash would have been expected to have a larger effect than observed on the viscosity;

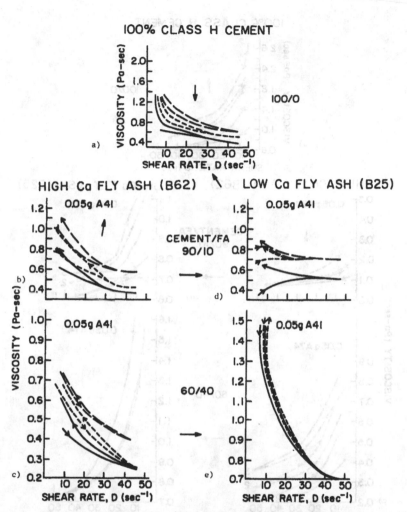

Figure 8. Rheological properties (measured on the Rheotron viscometer) of Class H cement (H-10) with different proportion of high calcium and low calcium fly ash mixed with salts of SNFC (A41) superplasticizer.

i.e., high-calcium fly ashes containing more reactive Ca-rich compounds would be expected to require more superplasticizer to generate comparable fluidity. However, the coarser particle size and irregularity of the particles in the low calcium fly ash rather than the chemistry controlled its rheological properties.

As higher Ca-content fly ash becomes more available, from sub-bituminous coals, it is necessary to know the controlling factors for obtaining the most economical pumpable mixtures. In this instance, particle size distribution and shape were the dominant factors.

Of the mixtures investigated, those with proportions of fly ash up to

104

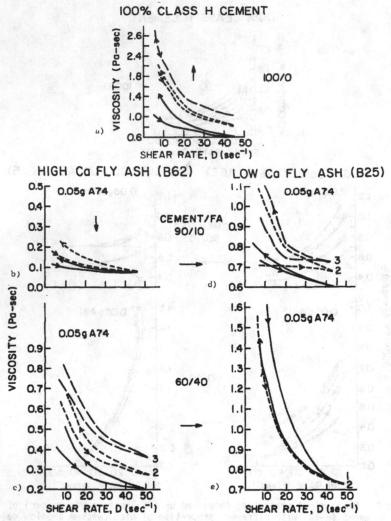

Figure 9. Rheological properties of Class H cement (H-10) with different proportion of high calcium and low calcium fly ash mixed with salts of SNFC (A74) superplasticizer.

about 40% are relevant for normal concrete mixtures, while the rheological properties of the high fly ash content mixtures (60 to 90%) are relevant for specialized applications such as in low-level nuclear waste management [6].

ACKNOWLEDGEMENTS

The material is based upon work supported by the National Science Foundation under Grants INT-8011922 and CPE-8112821. We thank many colleagues at the Central Research and Design Institute for Silicate Industry, in particular,

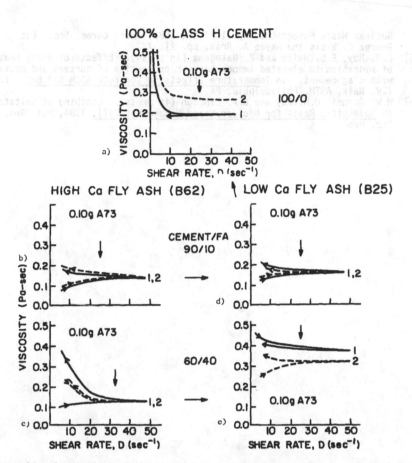

Figure 10. Rheological properties of Class H cement(H-10) with different proportions of high calcium and low calcium fly ash mixed with SMFC (A73) superplasticizer.

Mrs. Melindor Feher for her assistance in the research.

REFERENCES

[1] D.M. Roy and K. Asaga (1980). Rheological properties of cement mixes.
    V. The effects of time on viscometric properties of mixes containing
    superplasticizers; Conclusions. Cem. Concr. Res. 10 (3) 387-394.
[2] D.M. Roy, K. Luke and S. Diamond (This proceedings). Characterization of
    fly ash and its reactions in concrete.
[3] E.L. White and D.M. Roy (1983). Rheological behavior of cement slurries,
    mortars, and grouts. In Concrete Rheology, Proc. Symp. M, Mat. Res. Soc.
    (1982). Ed: Jan Skalny, pp. 108-119.
[4] Z. Nakagawa, D.M. Roy, and E.L. White (1984). Expansion and physical
    properties of MgO and SiO2-fume modified mortars. In Advances in Ceramics,
    Vol. 8, Nuclear Waste Management, Proc. 2nd Intl. Symp. Ceramics in

Nuclear Waste Management, April 24-27, 1983, Amer. Ceram. Soc. Ed:
George C. Wicks and Wayne A. Ross, pp. 710-722.

[5]   D.M. Roy, E.L. White and Z. Nakagawa (in press). Effects of early heat
      of hydration to elevated temperatureson properties of mortars and pastes
      with slag cement. In Temperature Effects in Concrete ASTM STP 858. Ed:
      T.R. Naik, ASTM, Philadelphia, PA.

[6]   M.W. Barnes, D.M. Roy and C.A. Langton (in press). Leaching of saltstone.
      In Scientific Basis for Nuclear Waste Management VIII, 1984, Mat. Res.
      Soc. Proc.

FLEXURAL STRENGTH AND FRACTURE PROPERTIES OF A FLY ASH BLENDED CEMENT

S. CHANDA AND J.E. BAILEY

Department of Materials Science and Engineering
University of Surrey   Guildford   Surrey   UK

(Received 29 November, 1984; accepted 8 February, 1985; Refereed)

ABSTRACT

The flexural strength, fracture toughness and Young's Modulus of an ordinary Portland cement plus 25 wt% fly ash paste (FACP) were compared with those of a control paste (HCP) at times ranging from one day to one year. At comparatively early times these properties develop more slowly in FACP. However, at comparatively late ages, e.g. ten months, FACP showed improvements over HCP for all of these properties. A complicating factor is that for both FACP and HCP, the flexural strength and toughness show maxima and the "crossover" times for the various properties differ. This paper analyses these results in terms of the observed fracture mechanism for HCP by taking into account the changes brought about in the microstructure of FACP, due to the incorporation of fly ash.

INTRODUCTION

Most studies on the mechanical properties of fly ash added cement paste (FACP) and concrete have concentrated on the compressive testing of fly ash added concrete. This is because the end use of this material is usually in compression. In addition, since the pozzolanic reaction is slower than the hydration of OPC, time and temperature are major considerations in this system.

A number of workers [1-4] have found that during the early stages of curing the compressive strength of fly ash added concrete (where the fly ash has partially replaced the OPC) is less than the corresponding control concrete at 20°C. At later stages the reverse is true. For example, Bennett [2] found the strength "crossover" point to be at 56 days, but this may vary according to the type of OPC and the type of fly ash used. In general, lower compressive strengths are encountered for fly ash added concrete for up to about three months of curing, and the strengths are greater after six months [5].

Work on the mechanical properties of cement pastes (as opposed to concrete) containing fly ash has been limited. It was decided to investigate the mechanical properties of the paste, and then the work could be extended to include the effect of adding aggregates. Furthermore, since compressive strength testing is affected by the geometry and the manner of force application [6], other more fundamental mechanical properties, such as fracture toughness, flexural strength and Young's modulus, were evaluated.

Previous work by Huang Shiyuan [7] on compressive and flexural strengths of fly ash added cement pastes, found similar crossover profiles of age vs. strength curves to those obtained by other studies involving concrete and mortar. However he noted that there was a reduction in both compressive and flexural strengths at later stages, and that this reduction was lessened by increasing the amounts of OPC

Table I. Oxide analysis of
the two cement powders.

Oxide	OPC	OPC + 25 wt % Fly Ash
SiO_2	20.1	28.4
I.R.	0.57	-
Al_2O_3	4.7	10.4
Fe_2O_3	2.5	4.2
Mn_2O_3	0.18	0.13
P_2O_5	0.12	0.15
TiO_2	0.22	0.41
CaO	64.4	48.6
MgO	2.1	2.0
SO_3	2.6	2.4
L.O.I.	1.4	1.3
K_2O	0.72	1.4
Na_2O	0.20	0.36
F	0.10	0.07
Free Lime	2.5	1.8

replacement by fly ash. (30% to 40% replacement gave rise to no
reduction in strength.) However no failure mechanism was proposed to
account for these results.

This paper presents the results of a comparative study of flexural
strength, toughness and stiffness of a fly ash blended OPC and the
control OPC paste (HCP). The development of these properties is
discussed in terms of possible and observed differences in the
development of the microstructure of the two materials.

EXPERIMENTAL DETAILS

Two batches of cement were provided by the sponsors of this
study. One batch was 100% OPC, and the other contained the same OPC
with interground fly ash (to give a 75 wt.% OPC + 25 wt.% fly ash
mix). Each batch was blended to ensure less variability between
different castings made from it. Then the batches were stored in a
number of sealed metal containers in order to retain the "freshness" of
the cement powder. The oxide composition of the two cement powders is
given in Table I. The specific surface areas of the OPC and the fly
ash were 405 and 397 m^2kg^{-1} respectively.

The castings were done at 0.3 water to solid ratio (w/s). The
cement was mixed with distilled water in a rotary motion planetary
mixer for ten minutes prior to casting in lubricated steel moulds. The
moulds were then vibrated for three minutes in order to remove large
bubbles. The cement was left to cure in a fog room (at ≈100% relative
humidity) for 24 hours. The blocks of cement were then demoulded and
placed under water for the required number of days. After curing the
blocks were cut into appropriate sizes using a circular diamond edged
saw. When necessary a notch was cut using a 0.15 mm thick diamond
edged wheel which gave a notch width of about 0.30 mm. All the cutting
was carried out under copious flow of water so as to prevent the
specimen from drying out. Great care was taken to keep the specimens
under water even during testing. It was ensured that the temperature
was kept to 20 ± 2°C throughout the casting and testing period. All
the tests were carried out after curing for 1, 3, 7, 28 days, 3, 6,
9 months and 1 year.

Flexural test specimens were tested using a three point bend rig in an "Instron 1195" machine. To compensate for any lack of parallelism in the specimen, the bottom two rollers were free to rotate slightly around the length axis of the specimen. Since the supports for the rollers were moveable, different specimen spans could be accommodated. A linear voltage differential transducer (LVDT), attached to the top roller support, was used to measure the deflection of the specimen. Unnotched specimens were used to determine flexural strengths and Young's moduli, while notched samples were used for stress intensity factor measurements. Fracture surfaces were examined using scanning electron microscopy.

RESULTS

Flexural Strength

The three point bend flexural strength specimens had dimensions of 80 x 25 x 14 mm. The cross-head speed of the Instron machine was 0.5 mm/min, and a span of 70 mm was used. The flexural strength, σ, was calculated using the standard beam theory formula:

$$\sigma = 3PL/2BD^2 \qquad \text{Eq. 1}$$

where σ is the failure stress, P the failure load, and L,B and D are the length, breadth, and the depth of the specimen respectively. The results are plotted in Fig. 1. It should be noted that the error bars in Fig. 1 and in later graphs, which were deduced from between 18 and 24 samples, represent 95% error bars for the estimated mean, μ. They have been calculated using the standard statistical formula:-

$$\mu = \overline{x} + t_c s/n \qquad \text{Eq. 2}$$

where $\overline{x}$ is the sample mean, s the sample standard deviation, n the sample size, and t_c is the appropriate percentage point of the t-distribution.

The flexural strength curves for both the HCP and the FACP (Fig. 1) were found to be similar to those obtained by Huang Shiyuan [7]. The FACP started with a lower flexural strength than the HCP; around 10 days it caught up with the HCP, and thereafter remained stronger. The remarkable feature of this graph was the peak at around one month followed by a small but noticeable reduction in strength for both the HCP and FACP. By about nine months to one year this reduction seemed to have levelled out, and the strength may even have started to rise. At the end of one year the FACP was found to be stronger in flexural strength than the control HCP by about 10%.

Young's Modulus

The stiffness or the Young's modulus, E, was determined from the three point bend tests. The Young's modulus was calculated from the following equation:

$$E = PL^3/4BD^3Y \qquad \text{Eq. 3}$$

where Y is the central displacement of the specimen.

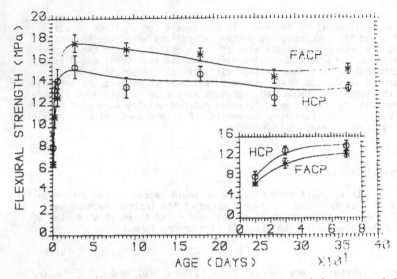

Fig. 1: Flexural strength (σ) vs. curing age of HCP [o] and FACP [*].

Initially the Young's modulus was determined using specimens of size 80 x 25 x 14 mm (and 70 mm span). Despite using the LVDT set-up, the values obtained for Young's modulus were about half of those quoted in the literature. After doing a series of different span to depth ratio tests it was found that a L/D of 30 or more gave reasonable results when compared with other measurements, and with the measurement of dynamic Young's modulus using a resonant frequency method. The cause for the initial inaccuracy in Young's modulus was thought to be due to shear strains being included in the specimen deflection measurement, as well as to crushing at the loading points of the specimen. Hence it was decided to change the Young's modulus specimen size to 160 x 9 x 5 mm (and 150 mm span). The results of the three point bend Young's modulus tests at different ages of curing are plotted in Fig. 2. Like the flexural strength curves, the Young's modulus for the FACP starts off at a lower value, and then has a higher value at later ages. The crossover point for the Young's modulus was found to be between three and six months. However there does not seem to be any evidence for any reduction in stiffness between 28 days and one year. It seems that between three months and one year there is only a small improvement in stiffness for the HCP, whereas the FACP keeps getting stiffer. By one year the FACP was found to be about 6% stiffer than the control HCP.

Stress Intensity Factor

One way to get an insight into the fracture mechanism of a material is to put a notch into a flexural strength specimen and then to test it. Assuming the material is homogeneous and linearly elastic, it is possible to calculate the toughness or the stress intensity factor, K_{IC}, using the following formula [8]:

$$K_{IC} = (3PL/2BD^2)C^{1/2}Y$$

Eq. 4

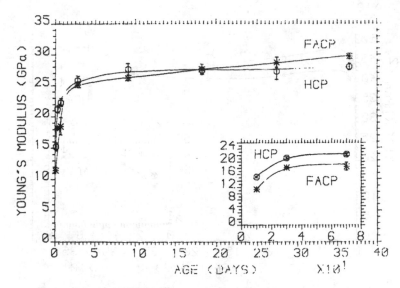

Fig. 2: Young's modulus (E) vs. curing age.

where C is the notch depth, and Y is geometrical factor and is a function of (C/D). The experimentally determined stress intensity factor was designated "$K_{IC}^{'}$" in order to avoid confusion with the real K_{IC} - this will be explained in the "Discussion" section. $K_{IC}^{'}$ was determined for both the pastes at the previously mentioned ages of curing.

The specimen size was 80 x 25 x 14 mm (and a 70 mm span), with a notch depth of 4 mm and a notch width of about 0.3 mm. The cross-head speed was 0.5 mm/min.

The measured stress intensity factors ($K_{IC}^{'}$) results are plotted in Fig. 3. The HCP values were similar to those quoted in the literature [9,10]. It can be seen that the HCP had a higher $K_{IC}^{'}$ than the FACP at early ages, and vice-versa at later ages, with the crossover at around ten months. There seems to be a peak for the HCP at around 28 days, followed by a drop of about 18% by one year. (Brown and Pomeroy [11] have also reported a drop in toughness of about 11% between 28 days and 84 days for a 0.35 w/s HCP using a notched bent beam test.) The FACP seemed to have a less obvious peak around one month, but after six months the toughness started to increase. By one year the FACP paste was ~10% tougher than the HCP using this K_{IC} test.

Scanning Electron Microscopy

Scanning electron microscopy was used to examined the fracture surfaces of the flexural strength specimens in order to investigate the fracture mechanism. After the specimen had fractured it was cut to a thickness of 3 to 5 mm using a diamond edged wheel machine. This

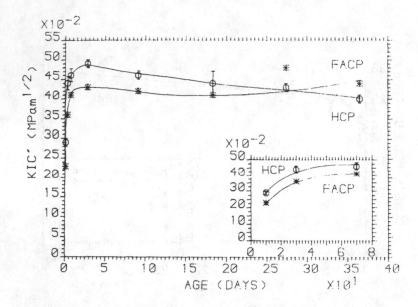

Fig. 3: Measured stress intensity factors (K_{IC}) vs. curing age.

facilitated the next step which was to dry the specimens in a vacuumed dessicator containing silica gel. The fracture surface was then coated with a thin layer of gold or platinum in a sputter coating device. The specimens were examined in a "Cambridge Stereoscan 100" SEM using the secondary electron imaging (SEI) mode.

A selection of the scanning electron micrographs are shown in Figs. 4-9. The first micrograph (Fig. 4) is of a one day old HPC paste. Most of the cement grains showed reaction products even at this early stage. These were mainly of the type I or type II varieties as classified by Diamond [12], i.e. small fibrillar particles emanating from the parent cement grains. The larger rods which can be seen in some areas of the micrograph were probably ettringite. There seems to be a fairly unreacted particle in the top left-hand corner of the micrograph. This was probably a belite (β-C_2S) particle which are known to hydrate slowly. As hydration proceeds the cement microstructure changes towards a denser one such as the one shown in Fig. 5. This specimen was fractured after nine months' curing. It consisted mainly of small particles of a somewhat indeterminate nature, which have been designated by Diamond as type III structures (see gel mass in the middle of Fig. 5). Of course less dense areas still remained, such as the areas towards the left edge and the right edge of Fig. 5. Here the fibrillar or foil-like morphology could not be superceded by the denser products due to large original inter-clinker granular distance.

The introduction of fly ash to the cement system caused a number of changes to the microstructure. First of all the 0.3 w/s FACP paste had a water/cement ratio of 0.4 compared to the 0.3 w/c of the 0.3 w/s HCP (if any adsorption of water on the fly ash particle is ignored).

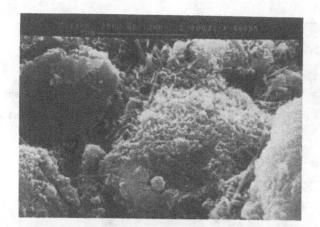

Fig. 4:
One day old
HCP

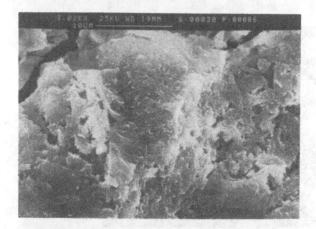

Fig. 5:
Nine month
old HCP

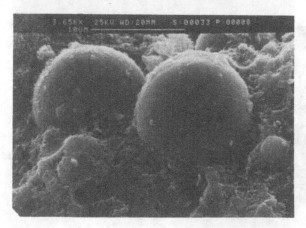

Fig. 6:
Fly ash
particles
embedded in
a seven day
old FACP

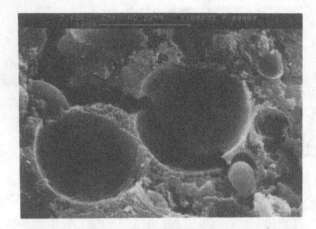

Fig. 7:
Holes left by
the fly ash
particles shown
in Fig. 6.
(Figs. 6 and 7
are matching
fracture
surfaces.)

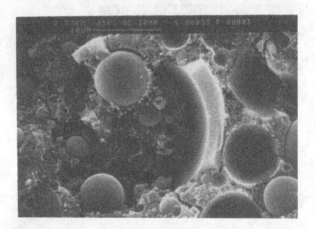

Fig. 8
Broken
plerosphere
embedded in a
seven day old
FACP

Fig. 9
Partially
reacted fly ash
particle in a
six month old
FACP

This was reflected in the slightly more open structure seen in the FACP. Figs. 6 and 7 are matching fracture surfaces of a seven day old FACP. The two fly ash particles in Fig. 6 were pulled out of the two holes shown in Fig. 7 during the fracture process. The coatings around the fly ash particles were very thin at this early age. The reaction products on the outside of the thin coating resembled the type I variety seen in the HCP. The fly ash particles seemed to be relatively unreacted and undiminished in size. At times it was possible to see broken plerospheres on the fracture surface as shown in Fig. 8. A part of the outer shell can be seen on the centre-right of the micrograph (the light feature running vertically). One of the fly ash particles embedded in the plerosphere was partially covered by type I fibrils. With an increase in curing age the pozzolanic reaction began to show its effect on the microstructure. Fig. 9 shows a six month old FACP. The coating around the fly ash particles had grown to about 1-2 μm in thickness. The fibrillar structure was seen even in this mature pastes. Another feature was the 0.5 μm or so thick gap between the reaction products and the fly ash particle. This has been observed also by Entin et al [13]. The other feature of interest in this micrograph is the needle-like structure on top of the exposed fly ash grain in the centre of Fig. 9. This is probably the outer skeleton of the unreacted micro-crystals of mullite (formed during the cooling process in the power station). Hence this may be another indication that the pozzolanic reaction was taking place by this age.

The other important microstructural feature in the Portland cement based systems is the calcium hydroxide (CH). These usually form as thin hexagonal plates which then lose their shape as massive amounts of CH are deposited in the microstructure. An example of the massive CH can be seen at the top left hand corner of Fig. 9. The CH is characterised by the relatively smooth, featureless surfaces.

DISCUSSION

The data obtained in this study provide evidence of reductions in strength and toughness. In view of the variabilities in strength results in materials of this kind, a note ought to be made about their statistical validity. It was decided to test 20 samples for each of the flexural strength determinations (as well as for the K_{IC} and E measurements). It was felt that this was a good compromise between time and accuracy, and the 95% error bars for the mean indicate the significance of the results.

The question then arises as to why there is an increase in flexural strength in both the cement pastes until about a month followed by a slight reduction. It is best to consider the microstructure of the HCP as a starting point. At relatively early times (e.g. a few days) the fibrillar microstructure was seen to predominate. Based on this type of fibrillar microstructure and on optical microscopy using diffuse illumination, Higgins and Bailey [10] have argued that a tied crack model operates in conventional cement pastes. The gain in strength up to 28 days can be thought of in terms of an increasing number of fibrils being formed with age as well as an increase in their size. The combined effect would be to increase the yield strength, σ_y, of the HCP, and hence increase the overall strength. An increase in toughness is also predicted by this model since a greater energy must be expended in order to part the larger number of enmeshed fibrils. Similarly Young's modulus will increase due to the increasing number of fibrils holding the structure together.

The reduction in flexural strength and toughness of HCP between 28 days and one year cannot be explained simply in terms of the fibrillar tied crack model envisaged for the earlier times. The fact that the Young's modulus increased by only a small amount between three months and one year suggests that the hydration occurs at a slow rate during this period. The values of K_{IC} plotted in Fig. 3 were computed on the assumption that the pastes were linearly elastic and homogeneous. However, the significance of the apparent change in measured K_{IC} must be assessed, because it is known that the apparent values for samples up to 28 days old vary with specimen dimensions. If due to microstructural changes the process zone (tied crack region) size varies then the meaningfulness of such changes is complex. Furthermore, a closer examination of the load/displacement graphs of the K_{IC} tests showed that the curves were not entirely linear. There was a deviation from linearity when around 88 to 98% of the final load was reached. A typical example is shown in Fig. 10 (inset). The ratio of the limit of proportionality (P_E) and the maximum load (P_F) was determined for each individual measurement. P_E/P_F is plotted against curing age in Fig. 10. The HCP shows a peak at around 28 days and the shape of the curve is somewhat similar to the K_{IC} vs. curing age curve in Fig. 3. The increase in P_E/P_F of HPC up to 28 days can be explained by the nature of the tied crack and reflects an increase in σ_y, the intrinsic strength.

The P_E/P_F for HCP drops betwen 28 days and one year. This result cannot be easily explained in terms of the tied crack model and it can be concluded that some phenomenon such as microcracking is occurring in the process zone. During this period there is no evidence of substantial changes in the microstructure and therefore the development of weak interfaces or internal microstresses must result from more subtle changes.

From the non-linearity of K_{IC} load/displacement graphs it is possible to conclude that the measured K_{IC} was an underestimation of the real K_{IC}. (Hence the measured values had been designated "K_{IC}".) It is possible that the real K_{IC} values did not change much between 28 days and one year since preliminary work on the work of fracture of HPC shows little change during this period [14]. Currently work is in progress to evaluate more realistic stress intensity factors.

Up to six months the replacement of 25% of the OPC by fly ash does not alter the general shapes of the respective age vs. flexural strength, stiffness and toughness curves. Clearly this is because the OPC part of the OPC and fly ash paste plays the active part in the mechanical properties build-up. However there were a few important differences. The flexural strength of the FACP was higher than the control OPC after about ten days. This is probably due to the improvement in rheology by incorporating the more spherical fly ash particles. Air bubbles, which act as strength reducing flaws, are thus reduced in size in the less viscous FACP during the casting procedure.

The rise in Young's modulus of the FACP was about 18% between 28 days and one year. Furthermore the stiffness vs. age curve for FACP (see Fig. 2) appears to be rising continuously even at one year. It would appear that the C-S-H, produced by the reaction between fly ash and CH and which replaces them, is more effective in contributing to stiffness. This may result from the existence of weak interfaces between the fly ash and the surrounding matrix which are progressively being removed. The effect of fly ash on the variation of toughness and

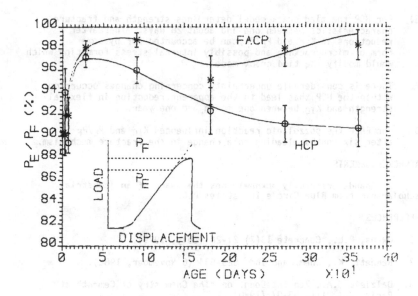

Fig. 10: Inset: typical load/displacement curve for stress intensity
          factor measurements.

          Main: P_E/P_F vs. curing age.

P_E/P_F with time begins to show up after around six months. The fly
ash and CH reaction would appear to lead to some competitive changes in
the microstructure which reverses the trend in K_{IC} and P_E/P_F.
The removal of CH may reduce the sites and interfaces available for
microcracking and internal stress generation. A detailed analysis of
these results will be reported elsewhere.

CONCLUSIONS

1. The replacement of 25% OPC by fly ash caused the paste to be
    weaker (in flexural strength), less stiff, and less tough at early
    times, and vice-versa at later ages. For example there was a 10%
    difference in flexural strength and a 6% difference in Young's
    modulus at one year.

2. The equilization points occurred any time between a few days and a
    few months depending on the particular property being studied.

3. Both HCP and FACP displayed flexural strength peaks at around 28
    days followed by a slight reduction.

4. The maximum value of flexural strength is greatest for FACP; it is
    considered that this is due to a reduction in the size of defects
    introduced into the material as a result of improved rheology of
    mixing.

5. Young's modulus shows an increase for FACP of 18% from one month
    to one year, whereas the HCP shows only a slight increase with
    time from three months onwards.

6. For HCP the tied crack model determines strength and fracture characteristics between one and about 28 days. Thereafter reductions in K_{IC} and P_E/P_F can be accounted for by some form of microcracking and possibly internal stress formation which would modify the tied crack model.

7. There is considerable uncertainty concerning changes occurring within the HCP that lead to the apparent reduction in flexural strength and K_{IC} between one month and one year.

8. For FACP the pozzolanic reaction influences K_{IC} and P_E/P_F after six months, leading to a change in the fracture mechanism.

ACKNOWLEDGEMENTS

S. Chanda gratefully acknowledges the award of an industrial scholarship from Blue Circle Industries PLC.

REFERENCES

1. Owens, P.L., Concrete 13(7) 21-26 (July, 1979).

2. Bennett, K., Chem. and Ind. 829-834 (6 November, 1982).

3. Dalziel, J.A., 7th Int.Cong. on "The Chemistry of Cement" at Paris, Vol. III, 93-97 (1980).

4. Owens, P.L. and Buttler, F.G., 7th Int.Cong. on "The Chemistry of Cement" at Paris, Vol. III, 60-65 (1980).

5. Berry, E.E. and Malhotra, V.M., ACI Journal 77(8), 59-73 (1980).

6. Kendall, K., Proc. R. Soc. Lond. A361, 245-263 (1978).

7. Huang Shiyuan, "Hydration of Fly Ash Cement and Microstructure of Fly Ash Cement Pastes", CBI Research No. 2 (1981), Swedish Cement & Concrete Research Institute.

8. Brown, W.F. and Srawley, J.E., ASTM STP 410, Philadelphia (1967).

9. Alford, N.McN., Groves, G.W. and Double, D.D., Cem. Concr. Res. 12, 349-358 (1982).

10. Higgins, D.D. and Bailey J.E., J.Mater.Sci. 11, 1995-2003 (1976).

11. Brown, J.H. and Pomeroy, C.D., Cem. Concr. Res. 3, 475-480 (1973).

12. Diamond, S., Proc.Conf. on "Hydraulic Cement Pastes: Their Structure and Properties", (Cem. Concr. Assoc., UK) at Sheffield, 2-30 (1976).

13. Entin, Z.B., Kuznetsova, T.V., Dmitriev, A.M. and Lepershenkova, G.G., Int.Symp. on "The Use of PFA in Concrete" at Leeds, 95-100 (1982).

14. Chanda, S. and Bailey, J.E., Unpublished work.

PROPERTIES AND POTENTIAL USES OF THE PRODUCTS RESULTING FROM THE FLUIDISED
BED COMBUSTION OF COAL WASHERY WASTES

DENIS G. MONTGOMERY
Department of Civil and Mining Engineering, University of Wollongong,
Wollongong, N.S.W., Australia

(Received 28 November, 1984; accepted 30 January, 1984; Refereed)

ABSTRACT

A fluidised bed combustion (FBC) process has been developed to a stage
that both coarse rejects and tailings from coal washery plants may be
successfully combusted.   During the combustion process excess heat is
produced, and the products from the process, spent rejects and primary and
secondary cyclone fines, are reduced in volume compared to the input
material.   Results of preliminary investigations into the properties of
the spent rejects and cyclone fines are reported.   A program has been
initiated to investigate and evaluate the potential uses of the FBC
process, including the use of the products in stabilised and unstabilised
forms in road construction, as a material suitable for the construction
of embankments, and in lean concrete mixes either as an aggregate or as
a pozzolan.   The characteristics of the FBC products are compared, where
appropriate, with those of the unburnt coal washery wastes.

INTRODUCTION

In 1980/81, Australian coal mines produced a total of 139 million
tonnes of raw coal, of which 106 million tonnes consisted of bituminous
and sub-bituminous coal.   Over 80 million tonnes of this coal were treated
by coal washing preparation plants to produce a higher grade product for
domestic and export markets. Coal treatment processes include crushing
and grading, washing to reduce the content of undesirable minerals which
will otherwise form ash, clinker or unwanted gases when the coal is used,
and blending to meet specifications.   The washing process results
in the production of waste material containing the unwanted minerals
and having high-ash content.   In 1980/81, 20.5 million tonnes of such
waste were produced.   The disposal of such large quantities of waste
material has been, and is, causing considerable concern. The environmental
problems of disposing of the wastes - including visual blight, the
production of silt and acid run-off, air-pollution, and the destruction
of local vegetation and wildlife - have reached proportions such that
State Government Departments are requiring certain mining companies
to find alternative methods of disposing of their washed fines-
traditionally stored as tailings ponds retained by embankments formed from
the coarse rejects of the washing process.

The C.S.I.R.O. has been investigating the fluidised-bed combustion
(FBC) process as a means for recovering energy from coal wastes and
providing an inert product suitable for safe disposal [1].   Since 1977,
a 4.5 MW (thermal) FBC pilot plant at the Glenlee coal preparation plant
at Clutha Development Ltd., near Camden, New South Wales [2] has been in
operation.   It is hoped that the process, which reduces the environmental
pollution resulting from the current disposal techniques, and provides
sufficient excess energy to make the process economically viable, will also
produce a range of products that can be used as engineering materials.
This paper outlines preliminary research into the possible utilisation of
products from the FBC bed combustion of coal washery wastes.

COAL WASHERY WASTES

Over decades of coal mining, more than 200 million tonnes of coal washery waste have accumulated in Australia. The coal washery wastes normally occur in two forms: (i) coarse rejects, coarse fractions of carbonaceous shale up to 200mm in size, often referred to as "coal wash", (ii) tailings, a suspension of fine (-0.5mm) clay, coal and other mineral matter in water at a solids concentration of 3 to 60%. Coal washery wastes show wide variations in composition. Older colliery spoil heaps contain sufficient quantities of coal that sometimes it becomes economically viable to re-wash the waste to recover a saleable coal. Spontaneous combustion can take place within the spoil heaps, and the use of the resultant "burnt colliery-shale" has had widespread use in Western Europe particularly in the United Kingdom [3].

The Australian Coal Industry Research Laboratories Ltd., has reviewed methods of waste disposal in Australian washeries [4]. All coarse rejects, amounting to 15.4 million tonnes in 1978/9, is disposed of in surface emplacements, either to form the walls of ponds or dams to contain the tailings or to recontour abandoned open-cut mine sites. Tailings, approximating 4.8 million tonnes on a dry basis, are discharged by most washeries for dewatering in ponds, dams, open-cut or abandoned underground workings. Tailings dams are the dominant method of disposal, accounting for 58% of tailings produced. In ponds and dams the solids are allowed to settle, while the water filters through the porous walls, comprised generally of coarse rejects, to be collected and recycled if possible to the washery. The tailings can be treated with mechanical dewatering devices to produce a material which can be blended with the coarse rejects prior to stockpiling, however little use of this approach has been made to date in Australia.

The most extensive and widespread use of coal washery rejects has been as landfill for projects including reclammation of mine-sites, playing-fields, parks, low-lying areas and back-filling of sand dunes [5]. The use of unburnt and burnt coal wastes is permitted in road construction subject to compliance with specifications [6]. The wastes can be used as bulk fill, sub-base or, if stabilised with cement, as a base material. In New South Wales, the use of coal washery wastes in road construction is not encouraged by the Department of Main Roads because of past problems relating to combustion, high clay content and grading. Extensive use of selected coarse rejects has been made on internal road construction at collieries and washeries in New South Wales: in public road construction of minor roads; in sub-divisions as bulk fill and as backfill to drains. Other uses of unburnt and burnt materials have been limited but include the manufacture of bricks, lightweight aggregate, cement manufacture, and as a soil conditioner.

FLUIDISED-BED COMBUSTION OF COAL WASHERY WASTES

For several decades attempts have been made to recover the energy content of coal washery waste material by burning in various types of boilers [7]. However, with the advent of the FBC system, the possibility of recovering available energy has been enhanced. FBC systems are commercially available for boiler applications and several demonstration plants for industrial boilers and power generation are operating [7]. Coal wastes have been successfully used as feed material for FBC plants in U.K. [8], West Germany [9], U.S.A. [10], China [11] and South Africa [12].

In a typical FB combustor, a bed of approximately 1mm diameter particles of inert material is held in suspension by fluidising air blown through the bed at velocities of 2-4 m/s. The fuel is normally crushed coal and the fluidising air supplies the oxygen necessary for combustion. A bed temperature of 800 to 950°C is maintained by balancing the heat released against that removed in the off gases or extracted by heat transfer surfaces in the bed. Heat recovered from the bed and the off-gases can be used to heat air or water, or to raise steam [13].

The 4.5 MW Glenlee plant used in our studies [14] is shown schematically in Fig.1. A substantial quantity of the energy content of the coal wastes can be recovered along with combustion products that may be suitable for use in engineering construction.

The plant was designed to operate using both coarse rejects and tailings feed material either separately or in combination. The emphasis of the work to date has been on utilising the tailings, which has supplied the main fuel input to the process, although a small amount of coarse reject material is often burnt as an auxiliary fuel. The coarse rejects are crushed to -12mm by a swing mill pulveriser and the tailings are thickened to a moisture content of about 40% (wet basis) before being fed

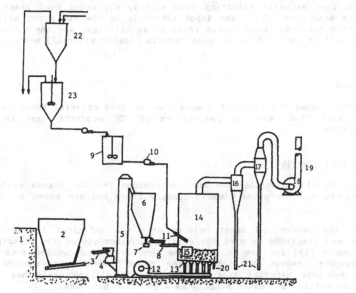

1	Loading ramp	12	Air blower
2	Main feed hopper	13	Air distributor system
3	Belt conveyer	14	Fluidised-bed combustor
4	Crusher	15	LPG burner
5	Bucket elevator	16	Roughing cyclone
6	Crushed-refuse feed bin	17	High-efficiency cyclones
7	Rotary table feeder	18	Induced-draught fan
8	Screw feeders	19	Stack
9	Tailings feed tank	20	Burnt rejects discharge
10	Metering pump	21	Cyclone fines outlets
11	Tailings injector	22	Primary tailings thickener
		23	Secondary tailings thickener

FIG. 1 SCHEMATIC DIAGRAM OF THE GLENLEE PILOT PLANT

to the combustor. The ash in the wastes is either drained from the bed as a coarse granular bed material up to 12mm in size or collected from the flue gases as fines below 0.5mm in size. The flue gases are treated by primary and secondary cyclones to collect the fine solid products.

It has been shown that the process can become profitable if heat is recovered for power generation [15]. Combustion trials performed on coal washery wastes from several sources have indicated that the recovery of energy can be successfully obtained by integrating a 5000 kg/h water-tube waste-heat boiler with the combustor [16]. A secondary benefit from the combustion process is that the volume of material requiring disposal is reduced to about two thirds that of the original waste.

CHARACTERISTICS OF FBC PRODUCTS

The initial investigations of material properties were confined to unburnt and burnt materials originating from the Glenlee Coal Washery and FBC pilot plant, respectively. C.S.I.R.O. have carried out combustion trials on feed material eminating from several different coal seams and respective washeries [16]. The input materials to the FBC plant will be subsequently referred to as coarse rejects and tailings, and the combusted products will be referred to as spent rejects (coarse material) and cyclone fines.

Composition

Table I shows the range of composition of feed material burnt in the Glenlee Pilot Plant and the composition of FBC products used in the investigation.

Particle Size Distribution

Particle size distributions were determined for the coarse rejects, spent rejects and the primary and secondary fines and are shown in Table II.

Both the coarse and spent rejects were too deficient in fine and medium gravel fractions to meet the grading requirements of the Department of Main Roads [17] for use in N.S.W. as densely graded crushed rock for road pavements, however both did meet the United Kingdom specification for granular sub-base material Type 2 [6]. Burnt colliery shale has been widely used within the United Kingdom as a granular sub-base.

TABLE I. COMPOSITION OF FEED MATERIAL TO AND PRODUCTS
FROM GLENLEE PILOT PLANT

	Coarse Rejects	Tailings	Spent Rejects	Cyclone Fines	
				Primary	Secondary
Moisture (%)	1.2-1.7	1.4-2.0	0.7	2.1	1.4
Ash (%)	61.8-76.0	20.6-40.2	99.0	91.2	94.0
Carbon (%)	9.8-21.9	38.7-54.3	0.3	6.1	4.4
Calorific Value(MJ/Kg)	4.3-9.5	19.2-26.8	-	-	-

TABLE II.   PARTICLE SIZE DISTRIBUTIONS OF COARSE REJECTS
AND FBC PRODUCTS FROM GLENLEE, N.S.W.

Sieve Size	% Passing by Weight			
	Coarse Rejects	F.B.C. Spent Rejects	Cyclone Fines	
			Primary	Secondary
13.2mm	97.5	100		
9.5	93.2	98.2		
4.75	70.9	83.7		
2.36	48.1	68.4		
1.18	23.1	43.6		
600 microns	8.0	14.2	100	
300	3.0	0.5	93	
150	1.5	0.1	59	
75	0.8	0	39	
52	0	0	31	100
6	0	0	0	97

Other Physical Tests

The main results of certain other physical tests carried out on the coarse and spent rejects are summarised in Table III.   Included in this table are results obtained by the Department of Main Roads on tailings supplied to Glenlee plant and tests carried out by Hoffman et al. [18] on tailings from washeries treating eastern bituminous coals in the U.S.A.

The non-cohesive nature of the spent rejects resulted in the Atterberg Limit Tests [19] being inappropriate for these materials.   Direct shear tests showed a lack of effective cohesion although the large angle of internal friction would indicate that stable emplacements could be constructed from this material.   Measurements taken with a constant head permeameter confirmed field experience that the coarse rejects can act as a good filter material.   The spent rejects were found to have a slightly lower permeability, but they have the advantage of not containing any soluble sulphates, thus indicating suitability for use as bulk fill.

POTENTIAL USES FOR FBC COAL WASHERY REJECTS

A list of the potential uses for the ash products from combustion, is reproduced after Waters [20] in Table IV.   It must be noted that, although the potential for various uses is available, it is most likely that the vast majority of the ashes would, in practice, be disposed of in emplacements similar to those in current practice.   This lack of full utilisation is in keeping with other waste materials such as blast furnace slag and fly ash. It has been estimated [21] that, in Australia, only about 10% of fly ash is sold for uses other than disposal by dumping.

This project was concerned with investigation of the potential use of FBC products in three civil engineering areas, namely: stabilisation, lean concretes and as filler material for bituminous concrete mixes.   As part of the investigation into the first two areas, the behaviour of the coarse rejects was determined in addition to that of the spent rejects.

Stabilisation of Coal Washery Wastes

Trial mixes were made to determine the effect of stabilising both the FBC spent rejects and coarse rejects (as received - 19mm) with either cement or a combination of cement plus fly ash (single source supply, Mummorah power station, N.S.W.; equivalent to Class F of ASTM C618).

TABLE III. PHYSICAL PROPERTIES OF COARSE AND SPENT REJECTS AND TAILINGS FROM GLENLEE
AND TAILINGS FROM SOME WASHERIES IN EASTERN U.S.A.

		Coarse Rejects	FBC Spent Rejects	Tailings* (Glenlee)	Tailings** (U.S.A.)	FBC Tailings
Atterberg Limits	Lower Liquid Limit (%)	25	30	22	–	27
	Lower Plastic Limit (%)	20	Undefined	Undefined	–	24
	Plasticity Index	5	Undefined	Undefined	–	3
Standard Proctor Compaction	Max Dry Density (kg/m^3)	1810	1670	1280	1030-1378	Undefined
	Optimum Moisture Content (%)	11	16	16.5	14.5-23.9	Undefined
Heavy Proctor Compaction	Max. Dry Density (kg/m^3)	2150	1810	–	–	–
	Optimum Moisture Content (%)	7	13	–	–	–
Maximum Dry Compressive Strength	(MPa)	0.9	Undefined	–	–	–
Direct Shear Test	Cohesion (C) (kPa)	0	0	–	0.1-2.3	0
	Angle of Shearing Resistance (φ)	43^0	54^0	–	8^0-53^0	45^0

*Results of tests carried out by Department of Main Roads [13].

**Results of tests on tailings from nine coal preparation plants in the U.S.A.,
after Hoffman et al [18].

TABLE IV. POTENTIAL USES OF ASH PRODUCTS FROM THE FLUIDISED-BED
COMBUSTION OF COAL WASHERY REJECTS [20]

Material	Potential Use	Substitute for
Burnt rejects		
Coarse aggregate (> 6mm)	High-grade fill and foundations Road construction	Coarse soft rock, gravel
Fine aggregate (2.0 - 0.6mm)	Fill, road pavements Building construction, as aggregate (lightweight) in concrete	Fine soft rock, gravel and sand
	Bricks, tiles, structural ceramics	clay, shale, sand
Cyclone fines (< 0.6mm)	Pozzolanic building material, as cement replacement or concrete additive Road construction	Power-station fly ash, fine sand
	Mineral filler in bituminous paving	Stone flour
	Soil stabiliser for road surfaces, embankments, and foundations	Lime, cement, fly ash
	Bricks, tiles, ceramics	Clay, shale, sand
	Miscellaneous: soil conditioner and other agricultural application; environment applications for oil spillage and liquid effluent treatment; filtration; Grouting and well cementing; source of alumina	Sand, Cement, fly ash. Clay Low-grade bauxite

TABLE V. MIX DETAILS FOR STABILISATION OF COAL WASHERY PRODUCTS

Mix Number	Added Moisture Content w/s	Stabilising agent content		Type of material (rejects)	Compressive Strength (MPa)	
		cement content	flyash content		7 days	28 days
1	6%	4%	-	Spent	0.7	0.9
2	10%	6%	-	Spent	0.8	1.0
3	10%	8%	-	Spent	0.6	0.9
4	10%	4%	4%	Spent	0.2	0.4
5	6%	4%	-	Coarse	1.7	1.9
6	6%	6%	-	Coarse	3.0	3.6
7	10%	8%	-	Coarse	3.7	5.0
8	6%	4%	4%	Coarse	2.5	4.2
9	6%	4%	2%	Coarse	2.2	2.5
10	6%	6%	3%	Coarse	2.9	4.6
11	6%	8%	4%	Coarse	4.3	6.8
12	6%	2%	4%	Coarse	1.3	1.7
13	6%	3%	6%	Coarse	2.5	3.7
14	6%	4%	8%	Coarse	2.8	4.0

Batches of cubes, of 150mm on a side, were made from each mix for subsequent testing in compression at ages of seven and twenty-eight days. The materials from each mix were batched by weight and thoroughly mixed in a rotary concrete mixer. The cubes were compacted in three layers, by external vibration through the use of a "Kanga" hammer which vibrated a steel plate on the free surface of the material within the cube mould. After compaction, the cubes were stored under moist conditions until tested in compression. Details of added moisture, cement and fly ash contents and respective compressive strengths for the stabilised mixes are presented in Table V.

In order to obtain mixes of suitable and comparable workability it was found to be necessary to incorporate sufficient added water to provide 10% moisture content when stabilising the spent reject material, whereas 6% additional moisture content was sufficient for the coarse reject material.

Stabilisation of spent rejects with cement or cement plus fly ash resulted in mixes of low compressive strength. Inspection of the cubes subsequent to testing indicated that the spent reject particles had a tendency to align in the direction of compaction, resulting in relatively little effective bond between adjacent particles. Compressive strengths were considerably higher for similar mixes incorporating coarse rejects. The addition of fly ash to these mixes resulted in improved strengths when the fly ash was used as a supplement to the cement content of the mixes and decreased strengths (7 and 28 days) when fly ash was used as a partial cement replacement.

Lean Concrete Mixes

Several mixes were made to investigate the behaviour of the spent rejects and coarse rejects in lean concrete mixes. Binder materials used were either cement or lime. Aggregate components of the various mixes included various combinations of spent and coarse rejects, fly ash, primary cyclone fines, 10mm basalt, 20mm basalt, 10mm blastfurnace slag and 20mm blastfurnace slag. Methods of batching, mixing, compaction, curing and testing were similar to those previously described for the stabilised mixes.

Details of the lean concrete mixes containing basalt aggregates are given in Table VI and mixes containing blastfurnace slag aggregates are given in Table VII. As evident in Table VI and Table VII for similar aggregate/cement ratios lean concrete mixes containing basalt, spent rejects and fly ash exhibit higher compressive strength than do the mixes containing spent rejects and basalt.

The use of fly ash in conjunction with the coarse rejects results in lower compressive strengths than when fly ash is not used. This is in contrast to the results for the stabilisation mixes given in Table V, where the addition of fly ash to the coarse rejects resulted in increases in compressive strength and the use of fly ash in conjunction with spent rejects was detrimental. The use of lime as a binding agent for mixes containing spent rejects resulted in mixes having little cohesion and no appreciable compressive strength.

For similar aggregate/cement ratios, in general, mixes containing basalt aggregate exhibited higher values of compressive strength than mixes containing blastfurnace slag. However, compressive strengths attained using slag aggregates were entirely acceptable for lean-concrete mixes, with the added benefit that all aggregates are comprised of industrial-waste materials. The use of fly ash as partial replacement of the spent rejects resulted in increased compressive strengths for the slag aggregate mixes.

The incorporation of primary cyclone fines (from the FBC process) in addition to a reduced fly ash content resulted in a lowering of the compressive strength compared to the similar mix containing full fly ash content.

Mix 30 was made for use in a test bed as a road base mix and subjected to repeated loading. After some 400,000 load cycles (1.0kN to 20.5kN) from a truck wheel and tire assembly the net deflection under the wheel was 20 percent of that obtained for a standard road pavement mix.

Cyclone Fines as a Filler for Bituminous Mixes

Cyclone fines from the FBC pilot plant were used as a filler material in standard 10mm bituminous asphaltic concrete mixes. Results have been reported elsewhere [22] and were shown to be comparable to those using conventional fillers. For use on heavily trafficked roads, analyses of the results indicated that a 5% FBC cyclone fines filler content at 5.5 to 6.0% bitumen content mix would be appropriate.

CONCLUSIONS

The fluidised bed combustion of coal washery wastes had been shown to be viable on an economic basis by recovery of energy in the form of heat produced during the process. The combustion process produces two main products, spent rejects and cyclone fines.

It has been shown that the spent rejects are non-cohesive in nature and cannot be successfully stabilised with either cement or a combination of cement plus fly ash, whereas the coarse washery rejects can be successfully stabilised with cement. The large angle of internal friction obtained for the spent rejects indicates that stable emplacements could be constructed from this material. Lean concrete mixes were made which incorporated various combinations of a selection of the following: spent

TABLE VI. MIX DETAILS FOR LEAN CONCRETES CONTAINING WASHERY REJECT PRODUCTS
AND BASALT AS AGGREGATES

Mix Number	Added Moisture Content (w/s)	Agg/ Cement Ratio	Type of Reject Material	Aggregate Percentages (by weight)				Compressive Strength (MPa)	
				Reject Material	Flyash	10mm Basalt	20mm Basalt	7 days	28 days
15	9%	8.5:1	SPENT	36.5	-	21.2	42.3	8.1	13.5
16	4%	9:1	SPENT	11.0	-	30.0	59.0	14.8	20.7
17	6%	9:1[a]	SPENT	55.5	-	14.5	30.0	4.5	8.1
18	6%	9:1	SPENT	55.5	-	14.5	30.0	14.5	16.5
19	6%	12:1	SPENT	57.0	-	14.0	29.0	6.3	8.6
20	9%	17:1	SPENT	36.5	-	21.2	42.3	2.2	4.6
21	9%	18:1	SPENT	34.4	5.6	20.0	40.0	6.4	10.5
22	9%	18:1	SPENT	31.7	8.3	20.0	40.0	6.5	12.1
23	9%	18:1	SPENT	34.4	1.4[b]	20.0	40.0	5.2	7.3
24	3%	19:1	SPENT	16.0	-	28.0	56.0	6.9	10.3
25	4.5%	19:1	SPENT	10.5	5.5	28.0	56.0	10.8	12.5
26	6%	19:1	SPENT	37.0	-	21.0	42.0	7.3	11.3
27	6%	19:1	SPENT	31.5	5.5	21.0	42.0	11.7	16.3
28	8%	19:1	SPENT	58.0	-	13.6	28.4	6.3	7.4
29	8%	19:1	SPENT	52.5	5.5	13.6	28.4	7.6	11.4
30	8%	25:1	SPENT	34.4	5.6	20.0	40.0	7.5	13.5
31	8%	9:1[c]	SPENT	55.5	-	14.5	30.0	0.1	No Result
32	6%	19:1[c]	SPENT	58.0	-	13.6	28.4	No Result	No Result
33	8%	19:1[c]	SPENT	52.5	5.5	13.6	28.4	0.05	1.0
34	7%	8.5:1	COARSE	36.5	-	21.2	42.3	15.0	20.3
35	7%	17:1	COARSE	36.5	-	21.2	42.3	5.2	7.4
36	9%	18:1	COARSE	34.4	5.6	20.0	40.0	2.1	3.2
37	9%	18:1	COARSE	31.7	8.3	20.0	40.0	2.7	5.1

NOTES: a = Mix includes 5% lime in addition to the 5% cement content
b = Mix includes 4.2% primary cyclone fines in addition to 1.4% flyash
c = Aggregate : lime ratio

rejects, unburnt coarse rejects, cement, lime, fly ash, primary cyclone
fines, basalt and blastfurnace slag. Lime was not successful as a binder
material for the lean concrete mixes. For mixes containing spent rejects
the highest strengths were obtained when fly ash was included in addition
to the cement content. No benefits were obtained with the use of fly ash
in mixes containing coarse rejects. Higher strengths were obtained for
mixes incorporating basalt aggregate although the strengths of the lean

TABLE VII. MIX DETAILS FOR LEAN CONCRETES CONTAINING WASHERY
SPENT REJECT PRODUCTS AND BLAST FURNACE SLAG AS AGGREGATES

Mix Number	Added Moisture Content (w/s)	Agg/ Cement Ratio	Aggregate Percentages (by weight)				Compressive Strength (MPa)	
			Reject Material	Flyash	10 mm Blast Furnace Slag	20 mm Blast Furnace Slag	7 days	28 days
38	4.5%	9:1	11.0	-	30.0	59.0	12.3	15.7
39	6%	9:1	33.4	-	22.2	44.4	18.8	19.7
40	8%	9:1	55.5	-	14.5	30.0	21.1	24.4
41	4%	19:1	16.0	-	28.0	56.0	6.8	8.2
42	5%	19:1	10.5	5.5	28.0	56.0	10.8	10.8
43	6%	19:1	37.0	-	21.0	42.0	5.9	9.0
44	6%	19:1	31.5	5.5	21.0	42.0	8.5	11.8
45	8%	19:1	58.0	-	13.6	28.4	4.5	6.4
46	8%	19:1	52.5	5.5	13.6	28.4	7.3	9.5

concrete mixes incorporating blastfurnace slag aggregate were quite acceptable. Cyclone fines can be successfully used as a filler for bituminous mixes.

ACKNOWLEDGMENTS

The experimental testing described was carried out by D. McQuart and N. Sutton. The author would like to thank Dr. G. Duffy of the Joint Coal Board for his assistance and advice.

REFERENCES

1. G.a.D. Szpindler, P.L. Waters and C.C. Young, Proc. 2nd Nat.Chem. Eng.Conf., Brisbane, 391 (1974).
2. R.D. LaNauze, G.J. Duffy and R. Sanderson, 8th Aust.Chem.Eng.Conf., Melbourne, 238 (1980).
3. P.T. Sherwood, TRRL Report 647, Transport and Road Research Laboratory, Dept. of Environment, Crowthorne, U.K. (1974).
4. N.D. Stockton, Aust.Coal Industries Research Laboratories Ltd., Report 09/1202 (1981).
5. R.W. Corkery, Geol.Surv. of N.S.W., Dept. of Mines, Report GS1976/309 (1976).
6. Ministry of Transport, Spec. for Roads and Bridgeworks, H.M.S.O., London (1969).
7. R.D. LaNauze, in Advanced Combustion Methods, Academic Press (1984).
8. A.A. Randell, D.W. Gauld, R.L. Dando and R.D. LaNauze, Proc.Eng.Foundation Conf., C.U.P., 286 (1979).
9. K.W. Belting, World Coal 5, 30 (1979).
10. I.G. Lutes and F.C. Wachtler, Proc.6th Int.Conf. on Fluidised Bed Combustion, Atlanta, Georgia, Vol.II, 405 (1980).
11. X.Y. Zhang, Proc.6th Int.Conf. on Fluidised Bed Combustion, Atlanta, Georgia, Vol.I, 36 (1980).
12. G.J. Duffy, Private communication (1984).
13. G.J. Duffy and J.W. Kable, C.S.I.RO., Division of Fossil Fuels, Invest. Report 142 (1984).
14. G.J. Duffy and J.W. Kable, Chem.Eng.Aust., Ch.E.4, 42 (1979).
15. P.L. Waters, Mine and Quarry Mechanisation, 184 (1976).
16. R.D. LaNauze and G.J. Duffy, A.M.I.C. Environmental Workshop (1984).
17. Department of Main Roads, M.R. Form 744 (1979).
18. D.C. Hoffman, R.W. Briggs and S.R. Michalski, U.S.A. E.P.A./D.O.E. Report E.P.A.-600/7-79-007 (1979).
19. K. Terzaghi and R.B. Peck, Soil Mechanics in Engineering Practice, Wiley (1962)
20. P.L. Waters, Inst. of Quarrying, Aust.Div. 15, 3, 19 (1975).
21. J. Beretka, Survey of Major Industrial Wastes and By-Products in Australia, Div.Building Research, C.S.I.R.O. (1978).
22. D. Pearson-Kirk and D.G. Montgomery, 2nd Aust.Conf. on Eng.Materials, Sydney, 261 (1981).

UTILIZATION OF FLY ASH IN ROADBED STABILIZATION: SOME EXAMPLES OF WESTERN U.S. EXPERIENCE

OSCAR E. MANZ, BRADLEY A. MANZ
  Coal By-Products Utilization Institute, Box 8115, University Station, University of North Dakota, Grand Forks, ND 58202

(Received 29 November; accepted 26 February, 1985; Refereed)

ABSTRACT

In contrast with pozzolanic bituminous fly ashes, the western American lignite and subbituminous fly ashes are both pozzolanic and cementitious. During the last three years, there has been much research and actual use of western ash alone and with lime or cement for roadbed stabilization. In contrast with lime stabilization which only applies to clay, the western ashes can be used to stabilize clay as well as silty and granular soils. Fly ash is moving from the experimental to standard alternative for pavement in many states. Some examples of western U.S. experience will be presented.

INTRODUCTION

By "western" fly ash we mean ash produced from combustion of low-rank western coals mined in North Dakota, Wyoming, and Montana. In describing the ash from coal combustion and conversion, the term "mineralogy" conventionally refers to the crystalline and noncrystalline phases present in the ash. In recent reviews of fly ash characterization, Diamond [1, 2] and McCarthy et al. [3] have pointed out that the high-calcium (Class C) fly ash produced from western lignite and subbituminous coals has higher CaO, MgO, and SO_3 and lower Al_2O_3 and SiO_2 contents than fly ash from bituminous coal (Class F). The higher CaO, MgO, and SO_3 contents of western fly ashes usually produce crystalline lime, periclase, calcium aluminate and calcium sulfate in addition to the dominant glassy phase. The mineralogy of Class C fly ash can play an important role in its behavior as a cement replacement because the lime, when hydrated, can activate a self-pozzolanic behavior in the ash itself and the two calcium phases can react with water in cementitious reactions such as ettringite formation [1, 2]. Indeed, one product marketed as "pozzocrete" (not trademarked) in the upper midwest has up to 80 percent fly ash replacement of cement in concrete.

The pozzolanic properties, low density, and good shear strength of fly ash have led to its use as a highway construction material in several areas of this country and in several European countries, particularly Great Britain. Fly ash can be substituted for many conventional materials which are dwindling in supply or escalating in cost. Use of this by-product also eliminates the cost and environmental problems associated with its disposal.

The pozzolanic properties of fly ash, which enable it to react with lime to form cementitious products, have made fly ash a good-quality base or subbase course material when used with lime or cement to stabilize aggregates and soils, or when used alone with lime or cement. Construction procedures utilize standard equipment and techniques for central mixing or mix in-place operations.

The American Society of Testing Materials (ASTM) has developed a specification for the use of fly ash with lime in lime-fly ash-soil aggregate mixtures [4]. This specification (ASTM C 593-Fly ash and other pozzolans for use with lime) establishes minimum unconfined compressive strength and durability requirements for mixtures using coarse-grained soil. These requirements are often specified for projects where lime and fly ash are used to stabilize fine-grained soils. The unconfined compressive strength criterion of 400 psi (2760 kPa) in seven days under accelerated curing conditions has proven to be quite acceptable, except that recommendations have been made for reducing this requirement to as low as 100 psi (690 kPa) for subbase applications [5]. In this procedure, a sample is held in vacuum for a specified time and then soaked. The unconfined compressive strength at the end of this test has been shown to correlate very well with the strength of samples which have been subjected to five or ten cycles of the freezing-thawing (but not brushed) [5]. The criterion specified for the strength at the end of the test has been suggested as 400 psi (2760 kPa), which allows for essentially no loss of compressive strength between the start and the end of the durability test. The advantage of the vacuum saturation test is that it can be performed in about an hour, whereas the standard durability test requires about 24 days to perform.

Fly ash can be used with lime, and, in some cases, alone, to modify subgrade soils to provide additional support for the pavement or to expedite construction. Lime-fly ash mixtures are able to reduce plasticity, improve drainage, and reduce shrinkage of many soils, as well as produce a cementitious matrix which further increases the soil's strength and durability. For subgrade modification purposes, no specific criteria have been established. Trial mixes are evaluated primarily on their success in modifying the soil properties in question. Mix-in-place construction procedures are most commonly used for stabilizing or modifying soils with lime and fly ash, although strict quality control is necessary.

Considerable research has been conducted on stabilization of soils with lime and Class "F" fly ash. While this information has been beneficial in assessing Class "C" fly ash stabilization of soils, it appears that more useful background information can be obtained from research and experience in the area of cement stabilization of soils. Our understanding of soil stabilization with self-cementing fly ash has further been hampered by our incomplete understanding of the chemistry of the self-cementing fly ashes and variations in fly ash properties between different sources. Of primary concern is the form of the calcium in the fly ash, because this dictates the soil-fly ash reactions that can occur. Current practice has been limited to evaluating the amount of a specified ash that is required to attain a desired change in physical characteristics of a specific soil. This could include changes in plasticity, shrink-swell potential, subgrade support capacity or shear strength. While for many applications this simple physical testing is sufficient, a more thorough understanding of the stabilization mechanisms will aid in expanding fly ash utilization.

Western Fly Ash Roadbed Stabilization Projects

Fly ash treatment of soils in the Kansas City area may be categorized into three general areas [6]. Fly ash is being utilized for drying soils to facilitate compaction, for treatment of soils to reduce shrink-swell potential and for stabilization to improve subgrade support capacity or shear strength. Fly ash stabilization of soils is used in construction of stabilized sections where an increase in shear strength of the soils is desired. These applications include stabilized base or subbase sections for pavements, stabilization of backfill to reduce lateral earth pressures and stabilization of embankments to improve slope stability. These applications generally require a more thorough understanding of the soil-fly ash reaction and more careful control of construction operations.

The Class "C" fly ash most commonly used for soil stabilization in the Kansas City area is obtained from the Kansas City Power and Light La Cygne Generating Station. All coal used at this facility comes from the Amax-BelAyre

Mine in the Powder River basin of Wyoming. X-ray diffraction analysis indicated that of the compounds present, approximately 10.6 percent by weight was tricalcium aluminate ($3CaO.Al_2O_3$) and 12.4 percent by weight was lime (CaO). The remainder of the calcium exists as calcium aluminum sulfate and calcium aluminum hydrate. This degree of crystallization of the calcium compounds appears to be unique to the La Cygne ash.

Referring to Figure 1, stabilization of a silty clay with only LaCygne fly ash did not produce very great strength. It was surprising that 12 percent fly ash produced greater strength than 15 percent.

Table I presents data on unconfined compressive strength of sand-fly ash (from Comanche station at Pueblo, Colorado) [7] at various ages using 10 percent and 15 percent fly ash. The strength gain between 28 days and 9 months is significant. With increasing time, the difference between 10 percent and 15 percent fly ash becomes small.

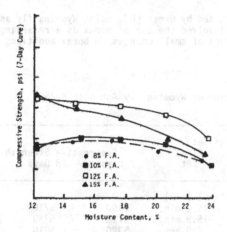

Figure 1 Compressive strength of a silty clay versus moisture content, for several levels of La Cygne fly ash additions.

TABLE I

Effect of Curing Time on Compressive Strength

10% Class C Fly Ash*		
7-Day	28-Day	282-Day
366	637	2003
366	597	2710
366 psi	617 psi	2356 psi

TABLE I, CONTINUED

Effect of Curing Time on Compressive Strength

15% Class C Fly Ash*		
7-Day	28-Day	282-Day
481	708	2175
505	764	2625
493 psi	736 psi	2400 psi

*Cured at 100°F in sealed containers.

Another study conducted by Grosz [8], using Wyoming fly ash from the Laramie River station, involved the use of borax as a retarding agent. In Table II, the large effect of small changes in borax additions is very evident.

TABLE II

Retardation of Wyoming Fly Ash With Borax

[1]Borax Content (%)	Time of Set		Compressive Strength (psi)		
	Initial Set	Final Set	7 Days	28 Days	56 Days
0	~3 (min)	~5 (min)	?	?	?
0.71	10.6 min.	15.2 min.	4490	5710	5840
1.42	58.9 min.	75.9 min.	5110	5730	6190
2.13	3.2 hrs.	4.0 hrs.	5320	5910	6650
2.84	6.7 hrs.	8.7 hrs.	4750	6020	6340
3.55	11.0 hrs.	14.7 hrs.	5540	6510	6890
4.26	17.6 hrs.	23.9 hrs.	5790	7130	7420

[1]Borax percent was based on weight of water. Water was 15-16% of the final mix by weight.

When it became evident that borax could alleviate to the problem of early hydration, the next step was to determine, in laboratory, if similar results could be obtained when Wyoming Grade "W" aggregate was used with varying percentages of fly ash. Unconfined compressive strength tests were performed using varying amounts of fly ash and a moderate percentage of borax. The results, Table III, were encouraging and gave a good indication of what was obtainable from soil stabilization with fly ash only.

TABLE III

Unconfined Compressive Strengths

[1]Fly Ash Content (%)	Compressive Strength (psi) Age Tested (Days)			
	2	7	28	56
10	150	150	100	180
15	380	320	350	420
20	660	840	1000	1130
30	1150	1520	1900	2770

[1]A 2.84% Borax solution was used for each mix

The use of fly ash-aggregate and lime-fly ash-aggregate stabilization using a retarder was reported by Koontz [9] for a Kansas secondary road project. Figure 2 shows the excellent results obtained by using HLFA-R 07 TM, a lignosulfonate retarder (obtained from the Walter N. Handy Company, St. Louis, Missouri), and the adverse effect of a fast setting fly ash on allowable compaction time.

North Dakota Fly Ash Stabilization Projects

Roadbuilding applications of lignite fly ash in North Dakota have used the largest tonnage to date. Poz-o-pac (not trademarked) road base mixtures, composed of 3/4-inch maximum aggregate, fly ash, and lime have been used for several years in the eastern half of the United States [10]. In 1971, the first lignite fly ash-lime-aggregate base (poz-o-pac) was placed in North Dakota near the site of the Basin Electric power plant. Approximately 13 percent fly ash and 2 percent hydrated lime were mixed with 85 percent aggregate and sufficient water to obtain maximum density. The laboratory procedure for determining the proper mix proportions is given in ASTM C593 (Specifications for Fly Ash and Other Pozzolans for Use With Lime [4]). Since 1971, two other North Dakota projects involving lignite fly ash poz-o-pac were completed.

A combination of 3 percent lignite fly ash and 3 percent lime has proven superior to a 6 percent addition of lime on a sub-grade A-7 soil stabilization project on I-29 in eastern North Dakota. We have conducted extensive performance tests involving stabilization of A-7 soils with lignite fly ash and lime [11]. Optimum strength was obtained with a 3 percent fly ash and 3 percent lime mixture, (Figure 3). A 15-mile portion used 7,000 tons of lignite fly ash at a savings of approximately $80,000 compared with use of lime.

A laboratory test program was undertaken by the North Dakota Highway Department [12] to determine the effects of the addition of lignite fly ash in lime-soil stabilization. The research program included testing of two cohesive and two relatively cohesionless soils in combination with lime and lignite fly ash in various quantities and proportions. Fly ashes from Otter Tail Power Company and Basin Electric Power Cooperative were used. Compressive strengths were determined for 65 different mixtures with lime content varied from 2 percent to 7 percent and lime to fly ash ratios ranging from 1:1 to 1:7. The effect of length of curing time of compressive strength test results and economic considerations were subjected to freeze-thaw and

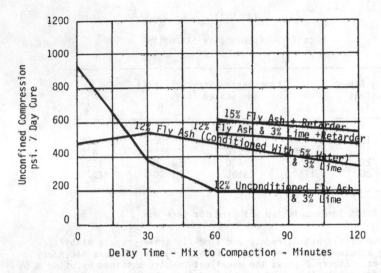

Figure 2. Strength versus compaction delay for fly ash stabilized base material with and without a lignosulfonate retarder.

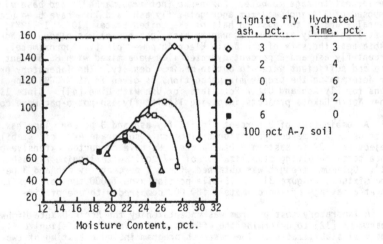

Figure 3. Compressive strength versus moisture content (7 days moist curing).

wet-dry durability testing. The results indicate that lignite fly ash in combination with hydrated lime can provide increased strength and durability for all of the soils tested.

There have been several lime-fly ash stabilized road projects constructed in North Dakota since 1981. The chemical and physical data for the various lignite fly ashes that were used are given in Table IV.

TABLE IV

Physical and Chemical Properties of North Dakota Lignite Fly Ashes Used for Stabilization

	Fly Ash Source					
	Heskett #83-405	Milton R. Young #82-244	Leland Olds #82-303	Coal Creek #83-275	Lewis and Clark #83-717	Antelope Valley #83-523
Chemical:						
Silicon dioxide (SiO_2)	16.8	25.0	43.9	45.6	34.2	72.2
Aluminum oxide (Al_2O_3)	9.6	10.3	14.0	15.5	18.3	9.7
Iron oxide (Fe_2O_3)	7.7	10.3	7.0	7.3	7.8	4.8
Sulfur trioxide (SO_3)	12.03	12.75	2.61	1.88	.88	14.62
Calcium oxide (CaO)	17.0	23.3	16.6	20.3	26.5	28.0
Magnesium oxide (MgO)	5.4	5.0	4.3	5.5	9.7	6.0
Sodium oxide (Na_2O)	7.0	7.9	7.9	1.0	0.8	7.6
Potassium oxide (K2O)	0.3	2.2	1.6	1.7	0.1	0.4
Loss on Ignition	23.27	2.3	0.28	0.14	0.34	2.91
Physical:						
Fineness						
% Retained on #325 sieve	49.32	18.24	29.6	16.13	20.43	6.41
Pozzolanic Activity Index with portland cement (%)						
Ratio to control @ 28 Days	44	88	90	106	78	81
with lime @ 7 days (psi)	290	1510	1030	1672	1450	1570
Water Requirement, % of control	112	93	88	92	93	91
Soundness						
Autoclave expansion (%)	0.12	0.21	0.23	0.09	0.26	0.13
Specific Gravity	7.18	2.72	2.48	2.50	2.72	2.70

In 1981, a 4-mile project was completed south of New Salem in Morton County. A ripper on a road grader was used to scarify an existing gravel road to a depth of 9 inches. Three and one-half percent hydrated lime and seven percent Heskett lignite fly ash were mixed with the proper amount of water by means of farm disc and road grader. Compaction was done with a sheeps' foot and pneumatic roller. An MC prime coat was applied, followed by a double chip seal. The cost of approximately $45,000 per mile was about half the cost of a hot mix mat. The laboratory testing was performed by the Coal By-Products Utilization Institute according to ASTM C-593, using a compressive strength criterion of 400 psi for the vacuum saturation test. With reference to Table V, even though the 400 psi requirements for vacuum saturation strength values were not achieved, strengths of 1550 and 3280 psi were observed after two years.

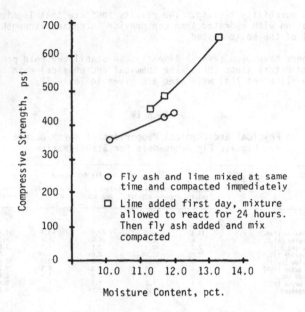

Figure 4. Effects of moisture, mixing, and curing sequence on lime-fly ash soil stabilization (3.5% hydrated lime, 7% Heskett fly ash).

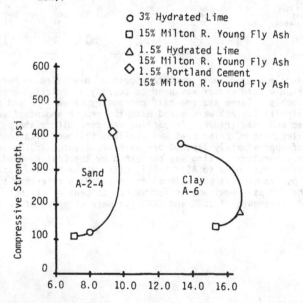

Figure 5. Comparison of lime, lime-fly ash, and cement-fly ash for stabilizing sand and clay.

TABLE V

Variation of Compressive Strength
With Curing Parameters

Composition

3/4" Gravel with clay, %	Hydrated Lime, %	Haskett Fly Ash, %	Moisture	Dry Density, pcf
89.5	3.5	7	11.4	120
89.5	3.5	7	9.8	118.4

Compressive Strength, psi

ASTM C 593 Vacuum Saturation	ASTM C 593 4-hr soak	7 Days, Sealed at 70°F	14 Days, Sealed at 70°F	28 Days, Sealed at 70°F	56 Days, Sealed at 70°F	90 Days, Sealed at 70°F	Cored From Stabilized Road After Two Years
817	313	246	315	431	553	834	1550
168	156	--	--	488	774	808	3280

During July 1983, a 1 mile stretch of road was stabilized using 3½ percent hydrated lime and 8 percent Heskett lignite fly ash. The previous road had a 4" hot mix pavement that was badly broken up and deteriorated. A single-pass pulverizer mixer was used to break up the hot mix mat and mix with 4 inches of clay and gravel below. Then the lime was spread and water added and the single pass mixer used again. After light rolling the road was left for 24 hours before the fly ash was spread and the single pass mixer used again. Then followed windrowing with a road grader back and forth across the road, with water added as needed. Compaction was begun using a sheeps foot and pneumatic rollers in sequence. Finally, the pneumatic roller was used alone to remove the sheeps foot marks. Immediately after final pass of the pneumatic roller, the MC-70 Prime was applied at 0.2 gals/sq. yd. A chip seal completed the road job.

When hydrated lime is used alone to stabilize clay soil, the lime is usually mixed in and the lime-soil mass allowed to mellow for a day or two before final compaction, to allow the lime to reduce the plasticity index. When dealing with western lignite and subbituminous fly ashes, any delay in achieving final compaction produces lower strength. Therefore, an investigation of moisture variation and mixing and curing sequence was conducted, (Figure 4). As shown, the highest strengths were obtained by mixing the lime first, allowing it to mellow for 24 hours, followed by the addition of fly ash and finally compaction. The lowest strengths were obtained when the lime and fly ash were added together and compacted immediately.

In 1982, sandy and clay soils near Center, ND, were tested using fly ash from the adjacent Milton R. Young Power Plant (see Table IV). Referring to Figure 5, a 3 percent hydrated lime addition was effective in stabilizing the clay but not the sand. Use of 15 percent fly ash without lime was not very effective with either the sand or clay, but the addition of 1 1/2 percent lime or cement with the sand-fly ash mix produced vacuum saturation strengths in excess of 400 psi.

Another study, Figure 6, involved vacuum saturation strengths for varying amounts of Leland Olds fly ash with and without hydrated lime [13]. Strengths are almost proportional to fly ash contents, with about 22 percent fly ash required to meet the 400 psi limit. It is interesting that 2 percent fly ash added to the 15 percent fly ash mixture increased the strength by 400 percent.

The importance of proper moisture content is demonstrated by Figure 7. The effect of moisture content on vacuum saturation strength is shown for variations of Coal Creek fly ash and hydrated lime. It is observed that with the lime constant at 1.5 percent, and with the fly ash increasing from 6 to 12 percent, the maximum strengths increased from 531 to 641 psi.

Although it is not usually economical to use 100 percent fly ash for a road job, there are occasions adjacent to power plants where its use is justified. Recently, the use of 100 percent Milton R. Young fly ash was investigated for possible use in a heavy equipment parking area. In Table VI, the effect of mixing sequence and moisture content, when only fly ash is used, indicates that change in moisture from 7.5 to 8.5 percent can change compressive strength by 130 psi. However, when using western ashes, a 6 hour delay in obtaining final compaction can result in drastically reduced strengths. Also, adding sufficient water to facilitate placing like concrete, reduces strength.

TABLE VI

Effect of Mixing Sequence and Moisture Content on
Compressive Strength When Using Only Fly Ash

Composition			
Minnkota Fly Ash, %	Moisture Content, %	Dry Density, pcf	Compressive Strentgh psi, 48 hcurs @ 120°F
100*	7.5	95	830
100*	8.5	99	960
100*	9.4	104	1100
100*	11.6	111	840
100*	12.3	113	830
100	22.8		320Δ
100*	12.5	83	160°
100*	13.5	80	60v

* 55# hammer, 12" drop, 3 layers, 25 blows/layer
Δ poured like concrete
° 6 hrs curing before compaction
v 24 hrs

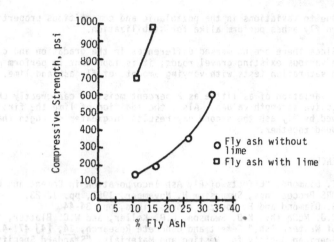

Figure 6   ASTM C593 vacuum saturation strengths for soil with varying
amounts of fly ash with and without 2-3% hydrated lime.

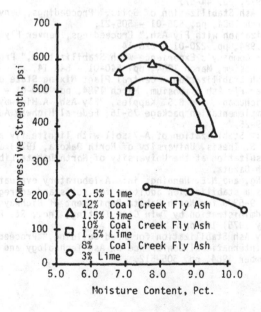

Figure 7   Effects of lime-fly ash variations and moisture content on
vacuum saturation strengths.

CONCLUSIONS

Due to variations in the pozzolanic and cementitious properties, not all western fly ashes perform alike for stabilization.

Since there are no marked differences in the gradation and clay content of the various existing gravel roads, it is important to perform ASTM C 593 vacuum saturation tests with varying amounts of fly ash and lime.

A variation of as little as 2 percent moisture can greatly change the compressive strength values. Also, the addition of lime the first day followed by fly ash the second day results in greater strength than when they are added together.

REFERENCES

1. S. Diamond, "Effects of Fly Ash Incorporation in Cement and Concrete," MRS Proceedings, Symposium N. November, 1981, pp. 12-23.
2   S. Diamond and F. Lopez - Flores, ibid, pp. 34-44.
3. G.J. McCarthy, K.D. Swanson, L.P. Keller, and W.C. Blatter, "Mineralogy of Western Ash," Cement and Concrete Research, 14, [4] 471-478 (1984).
4. American Society for Testing and Materials, "Standard Specifications for Fly Ash and Other Pozzolans for Use With Lime--Procedure C 593-76a," 1983 Annual Book of ASTM Standards, Vol. 04.02, 1983.
5. B.J. Dempsey, and M.R. Thompson, "Vacuum Saturation Method for Predicting Freeze-Thaw Durability of Stabilized Materials," Record No. 442, Highway Research Board, 1973, pp. 44-57.
6. G. Ferguson, "Fly Ash Stabilization of Soils," Proceedings, Denver Fly Ash Symposium, March 1984, pp. 505-01 - 505-21.
7. D. Smith, "Stabilization with Fly Ash," Proceedings, Denver Fly Ash Symposium, March 1984, pp. 220-01 - 220-14.
8. B. Grosz, "A Power Company's Experience with Stabilization," Proceedings, Denver Fly Ash Symposium, March 1984, pp, 540-01 - 540-14.
9. G. Koontz, "Fly Ash Stabilization - Central Plant Mixing State of Kansas," Proceedings, Denver Fly Ash Symposium, March 1984, pp. 580-01 - 580-11.
10. J.F. Heyers, R. Pichumani, and B.S. Kapples, "Fly Ash, A Highway Construction Material," Implementation package 76-16, Federal Highway Administration, 1976, pp. 19-25.
11. K.N. Derucher, in: Stabilization of A-7 soil with lignite fly ash and hydrated lime, (M.S. Thesis, University of North Dakota, 1973) 18 pp., available for consultation at the University of North Dakota Library, Grand Forks, North Dakota.
12. T.R. Dobie, S.Y, Ng, and N.E. Henning, in: A laboratory evaluation of lignite fly ash as a stabilization additive for soils and aggregates. (Final report No. 9-5373 for North Dakota Department of Highways and Federal highway Administration by Twin City Testing, Inc., St. Paul, Minnesota, January 1975) 145 pp.
13. O. Manz, "Lime-Fly Ash Stabilization for Road Building," Proceedings, Ash Tech '84, Second International Conference on Ash Technology and Marketing, London, UK, September 1984, pp. 505-512.

UTILIZATION OF FLY ASH IN OIL AND GAS
WELL CEMENTING APPLICATIONS

ASOK K. SARKAR
The Western Company of North America
P.O. Box 186, Fort Worth, Texas 76101

(Received 28 November, 1984; accepted 28 January, 1985; Refereed)

ABSTRACT

During cementing operations, the hydrostatic pressure of the cement slurry column is utilized to control the reservoir pressure. However, when the well extends through weak formations, it is often necessary to lighten the hydrostatic pressure of the slurry by lowering the density to avoid hydraulically fracturing the reservoir rock. Densities of cement slurries can be lowered by adding pozzolanic lightweight additives such as fly ash. Fly ashes have found utility in numerous oil and gas well cementing applications, namely: 1) fly ash can be dry-blended with regular Portland cements or hydrated lime for use in primary cementing; 2) lightweight cement compositions can be formulated by adding lime and finely ground quartz to fly ash/cement mixtures; and 3) an ultralight-weight pozzolanic microsphere cement, with densities as low as 0.96-1.64 g/cm^3, can be formulated by blending Portland cements with hollow microspheres isolated from fly ash.

This paper discusses the physical and chemical properties of each of these cement slurries, along with their relative advantages and disadvantages. Areas of future research are also discussed.

INTRODUCTION

In oil and gas well completion operations, steel pipe or casing is run into the well after the borehole is drilled. Once they are positioned downhole, they are cemented in place by filling the annulus between the borehole wall and the outside of the steel casing with a cement slurry. A typical well casing program is illustrated in Figure 1. When set, the solid impermeable cement sheath surrounding the casing prevents reservoir fluid movement between subterranean formations penetrated by the well. In addition to isolating oil, gas and water producing zones, the cement also helps in: 1) bonding and supporting the casing; 2) protecting the casing from corrosion; 3) preventing accidental blowout by quickly forming a seal; 4) protecting the casing from shock loads in subsequent drilling operations and 5) sealing lost circulation zones or thief zones.

The conventional method of cementing a well is to blend the slurry at the surface and pump it down through the casing, out the lower end of the casing and upward into the annulus surrounding the casing (Figure 1). The upward movement of the cement slurry through the annulus is normally continued until some of the cement slurry returns to the surface.

The downhole cementing process has grown in complexity since the early days of water flow control in mine shafts [1] in the 1880's and the cementing of steel pipe or casing in oil and gas wells as early as 1911 [2]. Today's oil and gas well cement slurry compositions are characterized by: 1) little or no filtration of cement slurry fluids (filtrate loss) into permeable formations; 2) little or no measurable waterseparation; 3) viscosities designed for optimum particle suspension, thus preventing solids settling; 4) optimum pumpability; 5) thickening times

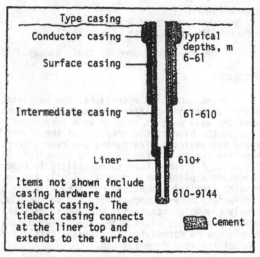

tailored to coincide with placement times; 6) high compressive strengths and 7) minimal shrinkage, strength retrogression or loss of permeability during the life of the well.

The cements to be used for this purpose, together with many of the additives that are included in the slurries, and the associated testing procedures are standardized in API Spec 10 [3].

SLURRY DENSITY

A very important consideration in cement slurry design is its density. The density of a slurry should always, except for remedial applications, be great enough to maintain well pressure control. The density of the slurry should also be higher than that of the drilling mud, which it must displace from the borehole. However, the hydrostatic pressure of the slurry should never exceed the critical formation pressure (fracturing pressure of the formation rock), otherwise, formation fracturing and excessive slurry loss may occur.

Neat cement slurries, when prepared from API Class cements using the recommended amount of water, will have a slurry density in excess of 1.8 g/cm^3 (15 lb/gal). In many deep wells, where high pressures are frequently encountered, cement slurries of higher density are often required. To increase the slurry density, additives are employed which: 1) have a specific gravity in the range of 4.5 to 5.0; 2) have a low added water requirement; 3) do not significantly reduce the strength of the cement; 4) have very little effect on the thickening time (length of time required for the slurry to reach an unpumpable consistency) of the cement and 5) are chemically inert and compatible with other additives. The most common materials used for increasing the density of cement slurries are shown in Table I. Of those listed, hematite has

Type casing
Conductor casing — Typical depths, m 6-61
Surface casing —
Intermediate casing — 61-610
Liner — 610+
Items not shown include casing hardware and tieback casing. The tieback casing connects at the liner top and extends to the surface. 610-9144

Cement

FIGURE 1
Typical Casing Program

Table I

Additives Used to Increase the
Density of Cement Slurries

Additive	Amount Used (percent by weight of cement)
Hematite	4 to 104
Ilmenite	5 to 100
Barite	10 to 108
Sand	5 to 25
Salt	5 to 18

been most widely used because it best meets the above criteria.

Conversely, if the formation cannot support a long cement column without fracturing, additives may have to be used to reduce the weight of the slurry. The advantage of using low-density cement slurries to achieve successful, competent cement jobs was not generally recognized until the 1940's. Since that time, lightweight slurries have been recommended for use in wells extending through weak subterranean formations that are very

Table II

Additives Used to Decrease the
Density of Cement Slurries

Additive	Amount Used (percent by weight of cement)
Bentonite	2 to 25
Attapulgite	0.25 to 2
Diatomaceous Earth	10 to 40
Solid Natural Hydrocarbon (gilsonite)	1 to 53
Coal	5 to 53
Walnut Shells	5 to 53
Expanded Perlite	5 to 21
Nitrogen	from 5 to 70*
Pozzolans (fly ash)	variable
Sodium Silicate	1 to 8

* Percent by volume of the slurry.

sensitive to hydrostatic pressures exerted by the column of cement. These additives also make the slurries more economical, increase their yield (volume of slurry per sack of cement) and sometimes lower the filtrate loss through permeation. The density of cement slurries can be reduced by either adding additional water, adding solids having a low specific gravity, or both. The materials commonly used to lower the density of cement slurries are shown in Table II.

POZZOLANIC ADDITIVES

Water and many of the additives used to lower the density of cement slurries contribute little or nothing to the ultimate strength development of the hardening cement. For this reason, pozzolans are preferred, because, being siliceous materials, in the presence of lime and water they develop cementitious properties. Pozzolans can be divided into natural and artificial pozzolans. The natural pozzolans are mostly of volcanic origin. The artificial pozzolans are mainly obtained by the heat treatment of natural materials such as clays, shales and certain siliceous rocks. Fly ash is a coal combustion by-product and is widely used in the oil and gas industry as an artificial pozzolan. This is the only pozzolan

144

covered by API and ASTM specifications. When Portland cement hydrates, calcium hydroxide is liberated. This chemical in itself contributes nothing to strength or water-tightness and can be removed by leaching. When fly ash is also present, its siliceous components combine with the liberated calcium hydroxide, contributing to both strength and water-tightness. Both ASTM Class F and Class C (high CaO) fly ashes are used.

Fly ash has a specific gravity of 2.3 to 2.7, depending upon the source, compared with 3.1 to 3.2 for Portland cements. When blended together, the difference in specific gravities results in a pozzolanic cement slurry of lower density than slurries of similar consistency made with Portland cement alone.

LOW DENSITY POZZOLANIC CEMENTS

Pozzolanic cement compositions are blends of API Class Portland cements and fly ash or simply a mixture of fly ash and hydrated lime. Of these two, the Portland cement and fly ash blend in more commonly used because of its superior strength development at low temperatures. By varying the amount of water added, a wide range of slurry densities can be prepared. Bentonite clay is often added to the slurries to limit water separation (free water). The density of these slurries can be decreased to as low as 1.44 g/cm^3 (12 lb/gal) by using bentonite and water, thus making the cement/fly ash blend slurries applicable in wells with bottomhole temperatures as low as 60°C (140°F). Conventional cement additives can be incorporated to produce predictable thickening times and compressive strengths. The lime/pozzolan reaction improves the resistance of the composition to sulfate attack by forming a dense microstructure which is impermeable to formation fluids. Typical thickening times and compressive strengths for different cement/fly ash (typically Class C) compositions are shown in Figures 2 and 3.

The blend of fly ash (typically Class F) and hydrated lime was developed for deep well cementing applications where the bottomhole temperature exceeds 93°C (200°F). These compositions often contain 2 wt% calcium chloride as an accelerator. Thickening time and compressive strength data for typical slurries are shown in Table III and Figure 4, respectively. Normally, no strength retrogression (loss of compressive strength with time) is experienced with this type of slurry when the amount of water is adjusted to yield a slurry density of 1.68 g/cm^3

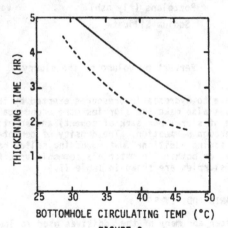

FIGURE 2
Typical Thickening Times of Pozzolanic Slurries Measured in a Pressurized Consistometer
—— (30% Class A Cement + 70% Fly Ash)
+ 2% Bentonite + 63% Water
---- (79% Class A Cement + 21% Fly Ash)
+ 2% Bentonite + 56% Water

(14 lb/gal). However, strength retrogression can be a problem with slurries having lower densities. Adequate strength retention is difficult to obtain with high temperature low density cement formulations derived from a mixture of fly ash and hydrated lime. The other disadvantage of this formulation is the very slow curing rates at relatively low temperatures, especially at the top of the cement column where the temperature can be as low as 60°C (140°F) and despite the presence of accelerators. In addition, the fly ash/lime slurries are not compatible with salt and are thus not well suited for cementing through salt zones.

Similar difficulties are also encountered with cement/ fly ash compositions, primarily because of their propensity for strength retrogression at temperatures exceeding 127°C (260°F). Even at lower temperatures, it is necessary to add substantial amounts of finely ground quartz to avoid this unwanted strength retrogression. Also, when the cement/fly ash mixture is used, the addition of bentonite as a free water control agent increases the viscosity of the slurry and often necessitates the use of dispersants to attain satisfactory pumpability.

Most of these disadvantages have been overcome by introducing to the industry cementing compositions which incorporate additional lime and silica flour to the cement/ fly ash mixture [4]. Like previously mentioned cementing compositions, these slurries can be prepared at varying densities, usually ranging from 1.38 g/cm³ (11.5 lb/gal) to 1.74 g/cm³ (14.5 lb/gal). Besides being compatible with most conventional additives used in the industry, these cementing compositions are also compatible with saturated salt solutions. The biggest single advantage of these complex cementing compositions is that they provide adequate compressive strengths at low temperatures. This makes them useful when cementing long strings of pipe in deep wells with high bottomhole temperatures in which

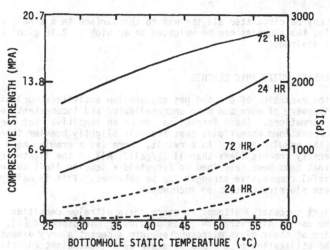

FIGURE 3

Typical Compressive Strengths of Pozzolanic Slurries
Cured at Atmospheric Pressure
—— (79% Class A Cement + 21% Fly Ash)
    + 2% Bentonite + 56% Water
---- (30% Class A Cement + 70% Fly Ash)
    + 2% Bentonite + 63% Water

Table III
Typical Thickening Times of Fly Ash/Lime Slurries
Measured in a Pressurized Consistometer

Slurry: Fly Ash + 25% Hydrated Lime + 2% $CaCl_2$

Water* (%)	Bentonite* (%)	Retarder* (%)	Bottomhole Temperature (°C)	Thickening Time (hr:min)
55	0	0.5	112	3:04
		0.75	112	6:27
		0.75	132	3:02
		1.00	132	6:26
		1.00	153	3:13
		1.50	175	4:38
79	2	0.5	112	4:53
		0.75	132	3:31
		1.00	153	5:09
		1.00	175	2:05
		1.50	175	5:09
114	4	0.5	112	3:04
		0.75	112	6:27
		0.75	132	3:02
		1.00	132	6:26
		1.00	153	3:13
		1.50	175	4:38

* Based on weight of fly ash

the slurry may be circulated all the way to the surface in a single stage. Additionally, the slurries can be weighted to as high as 2.16 g/cm^3 (18 lb/gal), if desired.

ULTRALOW DENSITY POZZOLANIC CEMENTS

With the expansion of oil and gas exploration activities worldwide came the discovery of more and more unconsolidated and incompetent (easily fractured) formations. These formations can be so sensitive that they will often break down when fluids that are only slightly heavier than water fill the annular space. As a result, there was a growing need for ultralow density cements (less than 11 lb/gal). All of the lightweight additives that have been described so far yield a density limit below which no useful compressive strength can be obtained. This is mainly because these slurries contain so much water.

The first successful attempt to reach these ultralow densities involved the use of high strength glass bubbles or microspheres [5]. The incorporation of these glass microspheres into a cement slurry essentially involves the utilization of encapsulated air as a lightweight additive. However, apart from being relatively expensive, these glass microspheres suffer from high water absorbency due to water permeation and bubble collapse, thereby changing their effective particle density under high pressures.

More recently, very small diameter microspheres (10-100 μm) have been successfully separated from the fly ash formed in coal burning power

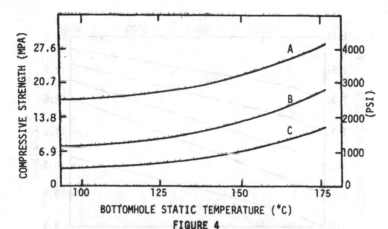

FIGURE 4

Typical 24 Hr Compressive Strengths of Fly Ash/Lime Slurries
Base Slurry: Fly Ash + 25% Hydrated Lime + 2% $CaCl_2$
A) Base + 55% Water*
B) Base + 2% Bentonite* + 79% Water*
C) Base + 4% Bentonite* + 114% Water*
*Based on weight of fly ash

plants. These pozzolanic microspheres are of relatively low cost and are strong enough to withstand mechanical shear, frictional forces, and hydrostatic pressure encountered during the blending and placement of cement slurries. This additive, because of its low density and relatively low water absorbency, has reduced the density limitations for pressurized cement slurries from 1.32 g/cm^3 (11 lb/gal) to densities less than the density of the mixing water [6]. The effective particle density of these pozzolanic microspheres ranges from 0.63 g/cm^3 at atmospheric pressure to approximately 1 g/cm^3 at 41.37 MPa (6000 psi). This low water absorbency coupled with their low density results in greatly reduced water requirements, as compared with other lightweight additives. The ultimate result is that the compressive strength development of a cementing slurry made with these pozzolanic microspheres will in many cases be much higher than other slurries of equivalent density.

At any given pressure, a certain percentage of the pozzolanic microspheres present in the slurry will fill up with water, thereby changing the overall slurry density. This is illustrated in Figure 5. Since the slurry density at the bottom of the cement column is of primary concern, the surface slurry density can be adjusted accordingly. Slurries prepared with pozzolanic microspheres behave essentially like the pozzolanic cement slurries and are compatible with conventional cement additives and saturated salt solutions. Thickening time and compressive strength data for typical slurries are shown in Figures 6 and 7.

These pozzolanic microsphere slurries have found applications in completing geothermal wells as well as oil and gas wells. The extremely high temperatures encountered in these wells mandate the use of cementing slurries with good thermal stabilities. Pozzolanic microsphere slurries not only develop higher compressive strengths and lower permeabilities than conventional high density silicate-extended slurries, but are far

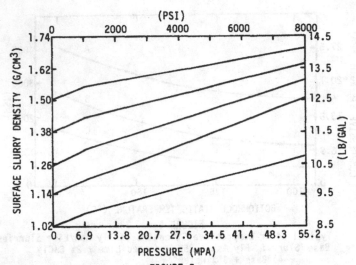

FIGURE 5
Effect of Pressure on the Density of Pozzolanic
Microsphere Slurries

superior in thermal insulating properties. These properties can substantially improve the thermal efficiencies of steam injection and geothermal wells.

FUTURE STUDIES

It is well recognized that the cement-fly ash interaction is influenced considerably by the chemical and mineralogical composition, as well as the fineness, of the two cementitious components. Initial cement hydration products, collectively called CSH* gel, convert to other crystalline forms as the temperature increases to 120°C (247°F). Normally, excess silica (35% by weight of cement) is added to maintain a low lime to silica (C/S) ratio so that no undesirable silicadeficient phases, particularly alpha dicalcium silicate hydrate (α-C_2SH), are formed. These phases are characterized by low strength and high permeability. Although the ideal soughtafter phase is tobermorite ($C_5S_6H_5$), this phase is converted to other high temperature phases at temperatures above 150°C (302°F).

The high temperature silica-rich phases are xonotlite (C_6S_6H), truscottite ($C_7S_{12}H_3$), sodium-substituted pectolite (NC_4S_6H) and carbonatecontaining scawtite ($C_7S_6\bar{C}H_2$). All of these phases have their inherent limitations, and for best strength and permeability retention characteristics, a suitable combination of all of these phases is desired [7].

---

* The following cement chemistry nomenclature is used: $C = CaO$, $S = SiO_2$, $H = H_2O$, $\bar{C} = CO_2$ and $N = Na_2O$.

Eilers et al. [7] have observed that adding powdered crystalline silica to cement slurries produces an excellent low permeability truscottite cement at elevated temperatures, whereas others have found that the use of fly ash as a source of silica leads to the production of reyerite, an undesirable form of truscottite [8].

The specific composition of fly ash additives, especially the form of silica, is thus very important, and there is an increasing need for their chemical and mineralogical classification. Studies of the kinetics of phase formations are also needed to help assure the long-term stability of fly ash cement compositions.

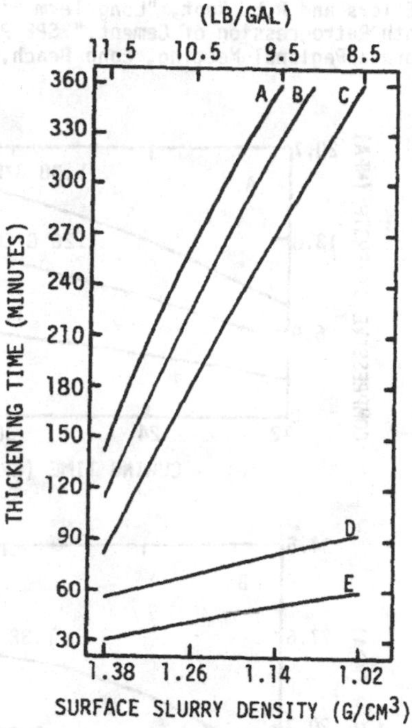

ACKNOWLEDGEMENTS

The author wishes to express his appreciation to the management of The Western Company of North America for permission to prepare and present this paper.

REFERENCES

FIGURE 6
Typical Thickening Times of Pozzolanic Microsphere Slurries Measured in a Pressurized Consistometer
Slurry: Class A Cement + Microspheres + 3% $CaCl_2$ + Water
Depths and API Casing Cementing Schedules: A) 310 m (1g-6); B) 610 m (2g-6); C) 1220 m (3g-6); D) 1830 m (4g-6); E) 2440 m (5g-6)

1. C. Dinoire, "The Application of Direct Cementation in Shaft Sinking," Trans., North of England Inst. of Mining and Mechanical Engineers, New York City, 31, 1-9 (1905-1906).
2. F.R. Tough, "Method of Shutting Off Water in Oil and Gas Wells," Bull., USBM, No. 163, 1918.
3. "Specification for Materials and Testing for Well Cements," American Petroleum Institute, Dallas, Texas, June 15, 1984.
4. A. Fincher and J.M. Goode, "A New Cementing Material for Deep Hot Holes," SPE Paper No. 5029 presented at Fall SPE Meeting in Houston, Texas, October 6-9, 1974.
5. R.G. Smith, C.A. Powers and T.A. Dobkins, "A New Ultra-Lightweight Cement With Super Strength," J. Pet. Tech., 1438-1444 (Aug., 1980).
6. W.M. Harms and D.L. Sutton, "UltralowDensity Cementing Operations," J. Pet. Tech., 61-69 (Jan., 1983).
7. L.H. Eilers, E.B. Nelson and L.K. Moran, "High-Temperature Cement Compositions - Pectolite, Scawtite, Truscottite, or Xonotlite: Which Do You Want?," J. Pet. Tech., 1373-1377 (July, 1983).

150

8. L.H. Eilers and R.L. Root, "Long-Term Effects of High Temperature on Strength Retrogression of Cement," SPE Paper No. 5871 presented at SPE California Regional Meeting, Long Beach, April 8-9, 1976.

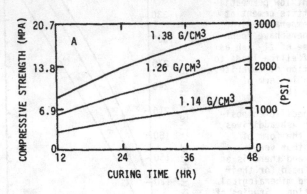

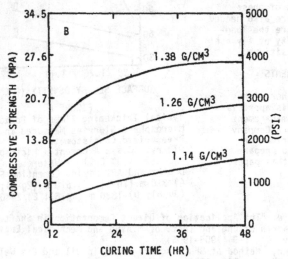

FIGURE 7
Typical Compressive Strengths of Pozzolanic
Microsphere Slurries
Slurry: Class A Cement + Microspheres +
3% $CaCl_2$ + Water
Curing Conditions: A) T=43.3°C, P=11.04 MPA
B) T=60°C, P=20.7 MPA

POTENTIAL RESOURCES FROM COAL FLY ASH

J. S. WATSON
Chemical Technology Division, Oak Ridge National Laboratory, P.O. Box X,
Oak Ridge, TN 37831

(Received 29 November, 1984; accepted 17 January, 1985; Refereed)

ABSTRACT

Recent studies at Oak Ridge National Laboratory (ORNL) and elsewhere
have identified various chemical processes for recovering useful materials,
such as alumina and iron oxides, from coal combustion fly ash. Based on
certain assumptions, each of these processes can yield useful products at
economical prices. Most processes leave a residual solid waste with a
volume only slightly less than that of the original waste. This residue
may not present a serious hazard, but its volume alone makes disposal
difficult for utilities with limited land available for disposal sites.
Characteristics of some of the residues are being studied to determine
possible beneficial uses or applications of these by-products.

INTRODUCTION

Ash from combustion of coal in electric utility boilers is one of the
major wastes produced by industry. These materials present significant disposal problems and expense, but they are also potentially useful materials.
The problems of ash disposal do not appear as urgent as they did a few years
ago when there was fear that some coal ash could be declared hazardous under
the Resource Conservation and Recovery Act (RCRA). The Environmental
Protection Agency relaxed connections between RCRA and drinking water
standards and provided at least a temporary exemption of coal combustion
wastes from a "hazardous" classification. Despite these developments,
disposal of coal ash, even as a nontoxic material, continues to become
increasingly more difficult and expensive because restrictions on ash disposal are increasing, and because suitable disposal sites are becoming more
difficult to find. Utilization of ash is expected to play an increasingly
important role in solving the disposal problems.

Coal ash can be divided into three components: slag, bottom ash, and
fly ash. The slag and bottom ash are collected at the bottom of the boiler
(either fused as in slag or granular as in bottom ash). Uses have been
found for much of these materials, but they constitute less than 30% of the
ash produced. The great bulk of the coal ash produced in large modern
utility boilers is fly ash, which consists of small particles blown from
the boiler and collected in the electrostatic precipitators or other off-gas
cleaning devices. Typical fly ash particles range in size up to 50 μm in
diam. The lower range of particle sizes is affected by the efficiency of
the collection devices; increased use of bag house filters may result in
larger quantities of smaller particles in the fly ash. Most fly ash particles are spherical in shape, indicating that they have fused during their
brief passage through the boiler, although irregular-shaped particles are
also found.

FLY ASH CHARACTERISTICS

Fly ash particles are formed from inorganic materials in the coal, and their composition varies both from particle to particle because of variations within a given batch of coal and from differences in coal sources. An excellent review of properties and characteristics of fly ash from many sources is given in a paper by Roy et al. in this volume. The internal structure of ash particles that is important to recovering metals was studied by Hulett et al. [1] after leaching with dilute solutions of hydrogen fluoride. Ashes from eastern U.S. coals contain crystalline mullite or quartz material surrounded by a more soluble amorphous silicate phase. The mullite crystals appear to have been growing during the brief passage of the particles through the boiler. Eastern U.S. coals usually have low concentrations of calcium and other alkali or alkaline earth elements, but those elements are more abundant in other coals found in the western United States. Kelmers et al. [2] showed that ash from coals with high calcium contents are less likely to form mullite and have more acid soluble phases.

POTENTIAL MINERAL RECOVERY METHODS

Since fly ash is a heterogeneous collection of particles, physical separations can yield particles of different size and/or composition with relatively little cost in energy or reagents. Magnetic separations can yield a useful product that is higher in iron and in density than the bulk ash. This product is a promising substitute for natural magnetite as a heavy medium for separation of mineral particles from coal [3,4]. The spherical shape of the particles from fly ash appears to enhance the usefulness of the material by lowering the viscosity of the heavy media slurry. The quantity of heavy magnetic material that can be obtained by magnetic separation differs considerably among ash sources, and some ashes offer little potential magnetic materials. The market for heavy media is limited, and one study [5] estimated that one to three magnetic separation facilities with the capacity of 40,000 tons of magnetite/year, the capacity of an existing plant [4] built by Halomet Corporation, would approximately meet the U.S. need for heavy media magnetite. However, increased cleaning by the coal industry could increase the demand by a significant factor.

A number of methods have been suggested for chemically separating metal oxides such as alumina from coal ash. Every approach focuses on recovery of the alumina in the ash because it has the highest value of any material with a large market. Other products have either lower total values (quantity times price) or insufficient market to absorb a significant fraction of the material in the ash. The choice of products and the available market for those products are important considerations in recovery of minerals from ash or any waste.

The best-known methods for recovering alumina from coal ash are the lime-sinter or lime-soda-sinter methods. These approaches are still being investigated at Iowa State University [6,7], and a commercial plant is reputed to have operated successfully in eastern Europe using a form of these processes. These processes involve dissolving the alumina in an alkaline solution and subsequently crystallizing the alumina. All of the nonalumina in the ash is incorporated into a second product, cement. The technology for this approach seems to be established, and the economics look promising [7], if the second product, cement, can be marketed at or near current cement prices. This is a major concern for these processes since they are likely to produce more cement than local markets can absorb.

Another promising approach involves chlorinating the ash under reducing conditions to volatile aluminum chloride [8,9]. This process is in an early stage of development, and several problems and details need to be worked out before it can be compared with other more developed processes. Separation of volatile impurities is one potentially serious problem which will require considerable development. If commercial plants for reducing aluminum chloride rather than oxide become available, the potential of this process for producing anhydrous aluminum chloride may become an advantage.

Two acid leaching approaches deserve mention, and one of these appears to be the most practical path to recovery of minerals from fly ash. The first is the Calsinter process [2,10] which, like the lime-sinter method noted previously, starts with sintering of the ash at approximately 1200°C with a calcium-containing material to break up insoluble mullite crystals in the ash. The sinter step is energy intensive, but this could be at least partially offset by the use and treatment of another electric utility waste, flue gas desulfurization waste (FGDW). The Calsinter process can produce a dry sulfate waste as well as alumina and iron oxide products. If sufficient credit is given for treatment of the sludge, the Calsinter process could be attractive. Any strong mineral acid probably could be used in the leach step, but sulfuric acid was used in experimental tests of the process since FGDW consists principally of sulfates.

The second acid extraction process is a simple direct leach of untreated ash with a strong mineral acid. This approach dissolves only the aluminum in the amorphous silicate phase, not that in mullite. As a result, the fraction of the total aluminum that can be recovered is low, perhaps less than 50%, depending upon the amorphous silicate contents. This process has been the subject of research by this author and co-workers at ORNL and will now be discussed in detail.

DIRECT ACID LEACH PROCESS

Process Flowsheet

A schematic flowsheet for an acid leach process, using approximately 6 to 8 $\underline{N}$ HCl, is shown in Fig. 1. Dissolution of the alumina requires approximately 2 h at reflux temperature (approximately 100°C). There is little incentive for using a more dilute acid since the leach liquor acid (chloride) content must be increased later for ion exchange and crystallization steps. The amount of residual solids is small, but no significant problems were found in settling and filtering the material. The residue must be thoroughly washed to remove excess HCl before the solids can be sent to waste disposal or to some practical use.

The fraction of the aluminum recovered depends on the composition of the ash and, to a lesser extent, on its temperature history in the boiler. Data for a number of ashes [2] are shown in Fig. 2. There is considerable scatter due to differences in the temperature histories of these different plants and perhaps due to specific differences in the composition of the inorganic matter in the coals. The composition of the leach solution from a typical ash (ASTM Class F) produced from an eastern U.S. coal is shown in Table I. This ash is from an East Tennessee coal burned at TVA's Kingston Power Plant. Approximately half of the aluminum is leached from the ash, and about two-thirds of the iron is leached. Aluminum and iron are the only major components in the leach solution (see Table II). Relatively large fractions of the calcium and a few other metals are leached, but none of these has high concentrations in the original ash.

154

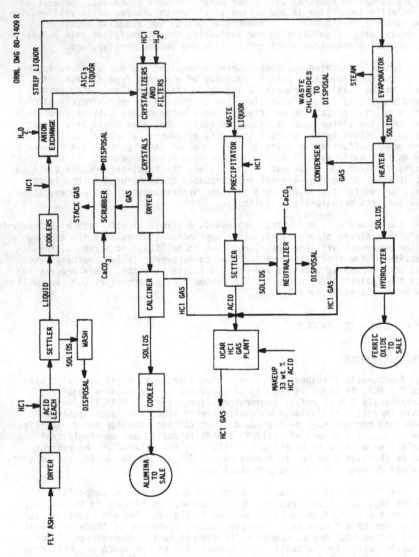

Fig. 1. Flow chart for HCl direct acid leach process.

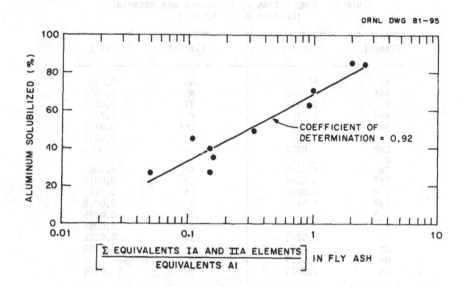

ORNL DWG 81-95

Fig. 2. Relationship between aluminum solubilization and base content of various ashes. Ashes were leached with excess HCl for 3 h at 17% solids at reflux temperatures. The coefficient of determination was 0.92.

Iron is removed first from the leach liquor by an extremely selective anion exchange method. First, the iron is oxidized to the +3 state by bubbling chlorine through the solution. Approximately two-thirds of the iron is in the ferric state before chlorine is introduced to the solution. The acid concentration is adjusted to give 10% free acid. The ferric iron can be removed by solvent extraction as suggested by studies of alumina recovery from clays [11] or by anion exchange. The latter approach was selected, and the selectivity of a strong base anion resin for ferric iron is illustrated in Table II. In the presence of high chloride concentrations, the ferric ions form anion complexes that load on anion resins. Table II gives the distribution coefficients for essentially all of the elements likely to be important to the iron product. The distribution coefficient for ferric iron is at least four orders of magnitude greater than that of any other element tested. Ferric chloride of essentially any desired purity can be obtained by this method. The few elements with slight affinity for the resin would be quickly displaced by the ferric iron and concentrate slightly in front of the iron-loading front moving down the ion exchange column. The purity of the iron product can be very high if one rejects (probably recycles) the iron at the very front of the loading wave.

The ion exchange is a cyclic or batch step, and one column can be eluted while the others are being loaded. Elution (removal) of the iron requires only washing the column with water or dilute HCl. Without suf- ficient chloride, the ferric chloride anion complex disappears, and the ferric cations are easily and quickly washed from the resin. This leaves the resin in the chloride form ready for reuse in the loading cycle.

Table I. Composition of reference ash material
(Eastern U.S. fly ash)

Element	wt %	Element	wt %
Al	15.2	Na	0.240
As[a]	0.0346	Nb	0.0016
B	0.0213	Ni	0.0114
Ba[a]	0.110	P	0.231
Be	0.0099	Pb[a]	0.0121
Ca	1.04	Sc	0.0031
Cd[a]	0.0002	Se[a]	0.0013
Ce	0.0175	Si	23.1
Co	0.0045	Sr	0.105
Cr[a]	0.0164	Th	0.0019
Cu[a]	0.0201	Ti	0.784
Fe[a]	7.84	U	0.0018
Hg[a]		V	0.0295
K	2.20	Y	0.0064
Li	0.0273	Zn[a]	0.0226
Mg	0.745	Zr	0.0203
Mn[a]	0.0195	-----	-----
Mo	0.0029	Ra[a]	5 pCi/g

[a]Listed in EPA and USPHS Interim Primary and
Proposed Secondary Drinking Water Standards.

Table II. Leachate composition after a 3-h leach with
8 N HCl at reflux temperature and 50% pulp density

Element	Resulting leachate concentration (g/L)	Element	Resulting leachate concentration (g/L)
Al	42.85	Mo	0.02
As[a]	0.12	Na	0.84
B	0.07	Nb	<0.01
Ba	0.07	Ni	0.04
Be	0.01	P	0.85
Ca	4.52	Pb[a]	0.04
Cd[a]	0.01	Sc	0.01
Ce	0.05	Sr	0.39
Co	0.02	Th	<0.01
Cr	0.05	Ti	1.22
Cu	0.08	U[a]	<0.01
Fe	35.24	V	0.11
K	6.81	Y	0.02
Li	0.10	Zn	0.09
Mg	2.50	Zr	<0.01
Mn	0.07		

[a]Calculated values.

The example flowsheet shown in Fig. 1 calls for hydrolysis of the ferric chloride to an oxide product [12]. The value of the product was estimated at the high side of an ore price, but the product may need to be pelletized for use in most ore reduction systems. A portion of the iron oxide may be sold at a value significantly above the ore price because of its high purity.

Aluminum chloride hexahydrate can be crystallized from the iron-free solution leaving the ion exchange columns. The crystallization can be initiated by thermal (evaporating water and excess acid) or concentration (salting out) driving forces. Initial studies have sparged the solution with HCl to raise the chloride concentration and precipitate the aluminum. Three stages of crystallization were suggested in the original flowsheet [13,14], but no more than two stages are now believed to be required. The crystallization and, especially, subsequent hydrolysis and calcining of the chloride to the oxide product are expensive steps in the flowsheet. Cost reductions in these operations would be especially helpful. Acid consumption must be minimized so that all the acid produced by hydrolysis of both the ferric chloride and aluminum chloride is recovered.

A small bleed stream is required to prevent buildup of impurities in the crystallization solutions. This stream consists principally of chlorides of calcium, potassium, sodium, and magnesium. It is desirable to recover as much acid as possible from this waste, because the acid lost to these chlorides is the major operating expense in the process, and discharge of soluble chlorides may not be acceptable. Wilder et al. [15] have considered adding sulfuric acid to the waste-bleed solution and recovering the HCl. As the resulting sulfate wastes are probably more environmentally acceptable, use of the sulfates has been suggested.

Economics

A preliminary cost analysis of such a system [14] indicated that the process could produce alumina at $275/ton and a high-purity iron oxide product at $60/ton without credit for detoxification of the ash (or no charge for the ash feed or for disposal of the waste products). The principal costs were for the hydrochloric acid consumed. With no charge for the ash, a reasonable assumption at this time, the process economics are not affected greatly by the fraction of the aluminum recovered.

The most recent assessment of the economics of mineral recovery from Kingston fly ash was completed by Kaiser Engineers (California) Corporation [15]. Laboratory support by ORNL was provided, and quotes and some testing were provided by potential vendors of the process equipment. This study looked into the flowsheet and economics with more detail and still found the concept promising and worthy of further study and consideration. An effort was also made to sell more products from the ash to enhance revenue. Both a base case and an "optimized" (more optimistic) case were studied, and the potential merit of multiple products was evident.

The design plant would process 1,000,000 tons of ash/year and produce a number of potential products, as shown in Table III. The ash feed would be a mixture of current ash produced by the power plant and ash dredged from disposal ponds. Although relatively few power plants will produce ash at this rate, two or more centrally located plants can provide ash at this rate, and numerous individual plants have enough ash in storage/disposal to meet this feed rate for decades. A separate steam plant included in this estimate provides heat and power to run the plant, and excess power is sold back to the utility. This arrangement separates the ash-processing plant from shutdowns in the power plant and permits the cost estimates to be based on load factors (operating time) that are independent of the load factor of any particular power plant. This is optional; one could use power and heat from existing boilers if it appeared more attractive to couple operations of the power plant and ash plants. No magnetic separation is included in this plant because the Kingston ash contains so little magnetic material.

Although the quantities and prices of these multiple products are con-jectural, they provide an estimate of the potential product value. Some products could flood existing markets, and average product prices could fall significantly below current levels. However, active marketing of these new materials, such as the leached ash, could find expanded markets, and average prices could meet or exceed these values. The lower alumina price used in this study reflects a less optimistic view of long-term alumina prices, and the higher iron oxide price is an attempt to value the high purities that can be achieved with the ion exchange separation. Of major importance is the value assigned to the leached ash. Because this is a new product, this value is conjectural, but there are several potential uses. The quantity is high enough that several uses may be needed to consume all of this material. Some uses may have even higher value and require some further treatment, but this is taken as a reasonable objective for the average value. Other products, such as gypsum and alkali sulfates, are desirable, but affect the overall economics less. Capital costs (with 25% contingency) were esti-mated to be approximately $244,000,000, and operating costs (excluding charges for capital) were $37,000,000. After reasonable charges for capital, the economics appear to be interesting and worth further study. If all of the suggested potential products were not marketed, some reduction in capital and operating costs may be possible. No credit was taken for reductions in ash disposal costs.

Table III. Optimized base case revenue tabulation -
EPRI metal oxides from fly ash study

Description	Units	Basis ($)	Amount produced	Annual revenue[a]
Total steam plant steam	lb/h		215,661	
Total cogenerated power	kW	@ stm. rt	16,741	
Total process power required	kW	0.045/kWh	10,131	3,829,500[b]
Total power sold to TVA	kW	0.035/kWh	6,610	1,943,300
Total power revenue	$			5,772,840[b]
Total process steam revenue	$	0.005/#/h		9,057,762[b]
Alumina sales revenue	tons/year	220/ton	158,100	34,782,000
Ferric oxide sales revenue	tons	400/ton	102,500	41,000,000
Gypsum sales revenue	tons/year	30/ton	45,800	1,374,000
Alkali sulfates sales revenue	tons/year	50/ton	81,000	4,050,000
Leached fly ash sales revenue	tons/year	50/ton	866,250	43,312,500
			Total revenue	$126,461,800

[a]Revenues indicated may vary significantly.
[b]Not included in total revenue.

OTHER POTENTIAL MINERAL VALUES IN FLY ASH

Early studies of waste processing focused on the recovery of alumina, a high-value product with a large market capable of absorbing all of the product likely to be produced from fly ash. Cement and iron oxide are promising products from some processes, and there has been interest in several minor components such as titanium and gallium. The most recent study of the economics of direct acid leach processes [15] noted the importance for marketing as much of the waste streams as possible. This offers additional revenues, but it also eliminates disposal problems and costs. The large volume of leached ash is particularly interesting because it is a very large waste stream, approximately 85% of the mass of the original ash. After leaching, this material is inert and not likely to present serious disposal problems except for its sheer bulk. If ash from several power plants were used for resource recovery, the wastes available would be large, and several large volume uses may be needed to consume all of the waste. Several potential uses have been suggested. Some have considerable value, but may require further processing to meet necessary standards and have a limited market. Large potential markets are usually associated with the construction industries, for example, as a component of concrete.

Particle sizes range from those of fly ash down to relatively small sizes with a significant fraction less than 5 µm in diam. The size distribution for residue from TVA Kingston ash is shown in Table IV. Note that a substantial fraction of the residue is less than 100 µm in diam. If certain sizes are desired for an application, size fractionation may be practical. The small size and the inert nature of the residue suggest that it could be used in a number of applications for filler materials. Further processing of the residue may produce useful materials with moderately high values. Separations may be needed to obtain the desired particle sizes and/or to remove (or recover) certain components. Microscopic examination indicates that there are different phases in the different particles. Thus, physical separations may yield purified and more useful products without large expenditures of energy or cost.

Table IV. Particle size distribution
in leached ash residue

Particle size d_i in µm	% less than d	$d_{i-1} < $ % $< d_i$
176	100.0	4.0
125	96.0	6.0
88	90.0	5.6
62	84.4	9.9
44	74.5	7.0
31	67.5	9.4
22	58.1	10.9
16	47.2	9.5
11	37.7	10.7
7.8	27.0	8.0
5.5	19.0	7.9
3.9	11.0	5.4
2.8	5.6	5.6

SUMMARY

Recovery and utilization of products from coal fly ash are potentially attractive options and deserve further study. Although all potential fly ash processing methods require further work and testing before they can become practical, all proposed approaches would benefit greatly by utilization of the residue, and a significant effort should be focused in that direction. The lime-soda-sinter process does not produce a solid waste stream directly, but failure to utilize all of the by-product in cement would result in a significant waste stream.

ACKNOWLEDGMENTS

This research was sponsored by the Electric Power Research Institute, Palo Alto, California, and by the Office of Fossil Energy, U.S. Department of Energy under contract DE-AC05-84OR21400 with Martin Marietta Energy Systems, Inc.

REFERENCES

1. L. D. Hulett, A. J. Weinberger, K. J. Northcutt, and M. Ferguson, "Chemical Species in Fly Ash from Coal Burning Power Plants," Science, 210, 1356 (1980).
2. A. D. Kelmers, R. M. Canon, B. Z. Egan, L. K. Felker, T. M. Gilliam, G. Jones, G. D. Owen, F. G. Seeley, and J. S. Watson, "Chemistry of the Direct Acid Leach, Calsinter, and Pressure Digestion—Acid Leach Methods for the Recovery of Aluminum from Fly Ash," Resources and Conservation, 9, 271 (1982).
3. N. K. Roy, M. J. Murtha, and G. Burnet, "Use of Magnetic Fraction of Fly Ash as a Heavy Medium Material in Coal Washing," in Proceedings of the Fifth Ash Utilization Symposium, Atlanta, Georgia, Feb. 25-27, 1979, p. 29.
4. R. G. Aldrich and W. J. Zacharias, "Fly Ash Magnetite--A Commercial Realization," in Proceedings of the Sixth International Ash Utilization Symposium, Reno, Nevada, DOE/METC/82-52, Vol. 2, 1982.
5. G. J. Kurgan, J. M. Balestrino, and P. S. Verma, "Market Survey of Fly-Ash-Derived Magnetite," Electric Power Research Institute Report No. EPRI CS-3615, 1984.
6. M. J. Murtha and G. Burnet, "Some Recent Developments in the Lime-Fly Ash Process for Alumina and Cement," Resources and Conservation, 9, 301 (1982).
7. J. A. Chesley, M. J. Murtha, and G. Burnet, "Utilization of Lime-Sinter Process Residue for the Manufacture of a Low-Alumina Portland Cement," presented at 96th Annual Meeting of the Iowa Academy of Science, Iowa City, Iowa, Apr. 27-28, 1984.
8. M. S. Dobbins and G. Burnet, "A Molten Salt Reactor for Dispersed Solid Phase Chlorination," presented at the AIChE Meeting, San Francisco, California, Nov. 25-30, 1984.
9. D. J. Adelman, "High-Temperature Chlorination of Coal Fly Ash," M.S. Thesis, Iowa State University (1980).
10. L. K. Felker and F. G. Seeley, "Chemical Development of the Calsinter Process for Recovering Resource Materials from Fly Ash," Oak Ridge National Laboratory Report No. ORNL/TM-7613, 1981.
11. J. A. Eisele, L. E. Schultze, D. J. Berinati, and D. J. Bauer, "Amine Extraction of Iron from Aluminum Chloride Leach Liquors," Report of Investigation 8188, U.S. Bureau of Mines, 1976.

12. J. W. Burtch, "The PORI Process: Regeneration of Hydrochloric Acid from Spent Pickle Liquor," Wire Journal, 57 (1976).
13. R. M. Canon, T. M. Gilliam, and J. S. Watson, "Evaluation of Potential Processes for Recovery of Metals from Coal Ash," Electric Power Research Institute Report No. EPRI CS-1992, Vols. 1 and 2, 1981.
14. T. M. Gilliam, R. M. Canon, B. Z. Egan, A. D. Kelmers, F. G. Seeley, and J. S. Watson, "Economic Metal Recovery from Fly Ash," Resources and Conservation, 9, 155 (1982).
15. R. F. Wilder, P. J. Barrett, L. W. Henslee, Jr., and D. Arpi, "Recovery of Metal Oxides from Fly Ash," Electric Power Research Institute Report No. EPRI CS-3544, Vols. 1, 2, and 3, 1984.

PART III

Coal Gasification Ash

CHARACTERIZATION OF A LIGNITE ASH FROM THE METC GASIFIER

I. MINERALOGY

GREGORY J. McCARTHY*, LINDSAY P. KELLER*, ROBERT J. STEVENSON**, KEVIN C. GALBREATH* AND AARON L. STEINWAND*
*Department of Chemistry, North Dakota State University, Fargo, ND 58105
**Natural Materials Analytical Laboratory, University of North Dakota, Grand Forks, ND 58202

(Received 28 November, 1984; accepted 13 February, 1985; Refereed)

ABSTRACT

Utilization or disposal of gasification ash requires detailed characterization of its chemistry and phase formation (mineralogy). A North Dakota lignite ash produced in the Morgantown Energy Technology Center (METC) gasifier has been studied in detail by x-ray diffraction and electron microprobe analysis. The ash was coarse (84% of grains larger than 1.0 mm) but a typical grain was composed of a dozen or more crystalline phases with dimensions on the micrometer scale as well as less abundant glass phases. Hard centimeter-size clinkers suggested partial melting followed by crystallization. Silicates (dicalcium silicates (C_2S), merwinite, Ca-Na-silicate (CNS), quartz), aluminosilicates (melilite, nepheline, carnegieite), oxides (ferrite spinels, periclase, hematite), calcite and minor zeolites comprised the dominant mineralogy. Microprobe analyses were obtained for large numbers of grains of the C_2S phases, CNS, merwinite, melilite, ferrite spinels and calcite. The remaining phases had crystal sizes too small for analysis. A model is proposed for the genesis of this ash based on the inorganic constituents of lignite and the gasifier operating conditions.

INTRODUCTION

December 1984 marked the full-scale start-up of the first commercial-scale coal gasification in the United States. Great Plains Gasification Associates (GPGA) of Beulah, North Dakota, began converting lignite coal into natural gas and other marketable by-products such as ammonia and naphtha. Ash is one by-product that is presently not sold. The ash is instead buried in clay-lined pits on the adjacent lignite mine site. The costs of disposal would be lowered or eliminated if this ash could be utilized in construction or as a raw material for manufactured products. Efforts are underway to evaluate existing utilization options and develop new uses for gasification ash [1,2]. These efforts are being made as generic as possible because success here will carry over into future gasification plants and would make coal gasification technology more cost effective. Proper disposal of the ash requires a thorough understanding of the nature of the ash and its behavior under the various physical and chemical conditions in the disposal site. Much of the work necessary to characterize the ash will also be useful in predicting ash behavior during by-product fabrication.

McCarthy et al. [3] and Stevenson [4] have described the general mineralogy and crystal chemistry of one lignite ash produced in a fixed bed gasifier. We report here on a more extensive study of another lignite ash provided for this project by the Department of Energy facility at the Morgantown Energy Technology Center (METC), West Virginia, in late 1983. The results will be given in three parts, with this paper being an x-ray diffraction (XRD) and electron microprobe study of the ash mineralogy. Stevenson and Larson [5] will describe the microscopic morphologies and compositions of the phases that make up the ash and Hassett et al. [6] will report ash leaching tests and provide correlations of these results with the composition and mineralogy of the ash.

LIGNITE GASIFICATION

The METC gasifier is a pressurized, stirred, fixed-bed, dry ash producer of the Wellman-Galusha type. Lignite from the Beulah-Zap seams was obtained from the Indianhead Mine, located south of Beulah, North Dakota. The coal, sized to 2 in. (5 cm) or smaller chunks, is fed into a lockhopper and then into the gasifier. The lignite is dried and then devolatilized as it descends against the counterflow of hot gases into the hotter portions of the gasifier. The coal then passes into the gasification zone where it reacts with steam to form H_2 and CO, and with O_2 to form CO and CO_2. Heat from the carbon combustion reaction supplies energy to the endothermic carbon-steam gasification reaction. An approximately 20 in. (50 cm) thick bed of ash sits atop a rotating grate. The mixture of about 2.5 parts of H_2O to one O_2 is injected below the ash grate. For the lignite run, the gasifier was operated at pressures of 100-200 psi (0.7-1.3 MPa).

To date, METC personnel have not found an effective method of measuring representative temperatures in the hot zones of the gasifier. However, estimates can be made from the appearance of the ash and the fusion temperatures derived for the specific ash composition [7]. Based on the presence of some large partially melted agglomerates, we estimate that the average bed temperature in the hottest part of the gasifier was 1050-1100°C (1920-2010°F) with temperatures as high as 1250°C (2280°F) in some areas for short periods of time. Estimated residence time in the gasification zone was 10-15 minutes. It should be noted that the ash cooling atop the grate prior to discharge was exposed to the steam/O_2 atmosphere. Water was also used to lubricate and flush seals between the ash pit and hopper. Thus, the possibility exists for formation of hydrated phases in the ash. The ash was discharged directly from the hopper onto a conveyor belt and then into two large steel barrels for shipment.

SPECIMEN PREPARATION

The ash sent by METC was light gray in color and had a consistency ranging from fine powder to agglomerates up to one cm in cross-section. The larger agglomerates were very hard and suggested to us that some melting of the ash had occurred during gasification. There appeared to be little or no unburned coal or char in the ash.

Ash specimens were prepared in the following manner. The contents of each barrel sent by METC were homogenized and representative samples of the ash were collected into a number of two liter containers. The contents of two of these jars were ground to -20 mesh. This product, here termed "bulk ground ash," was analyzed by atomic absorption spectrometry (AAS) for major elements (Na, Mg, Al, Si, K, Ca, Ti, Fe, S) and by x-ray fluorescence spectrometry (XRF) for selected minor elements. Three additional jars were separated by grain size using a series of standard sieves on a shaker. The size distribution of the ash is shown in Table I. The ash is relatively coarse with 84 wt.% of the agglomerates larger than 1.0 mm.

Aliquots of each size fraction were ground in an agate mortar. Additional specimens, rich in a small number of phases, were hand-picked with the aid of a binocular microscope. This procedure is essential to confirming phase identifications in the diffractograms of the bulk ground ash and ash fractions. The -20 mesh bulk ground ash was ground further in a Spex Mixer

TABLE I

Size Distribution of METC Gasificatication Ash

Sieve Number	Seive # Size	Wt. (g)	Wt. %	Sieve Number	Seive Size	Wt. (g)	Wt. %
-	>9.52mm	656	11.2	60	250–500μm	232	4.0
7	2.8–9.52mm	3142	53.8	120	125–250μm	129	2.2
10	2.0–2.8mm	446	7.6	230	63–125μm	95	1.6
18	1.0–2.0mm	664	11.4	325	45–63μm	46	0.8
25	710um–1mm	205	3.5	<325	<45μm	84	1.4
35	500–710μm	139	2.4		Totals:	5838	100.0

Mill using a steel jar and balls. Aliquots of this ash were mounted on glass microscope slides. Diffraction data were obtained using Philips manual or automated diffractometers, each equipped with copper tube, theta-compensating slit, diffracted beam monochromator, proportional (manual) or scintillation (automated) detector and solid state electronics. Phases were identified with the JCPDS Mineral and Inorganic Powder Diffraction Files and by comparison to the patterns of reference solid solution phases prepared in our laboratories. Crystal data were obtained for some of the solid solution phases as further verification that the phase had been properly identified. Using grains that were enriched in a particular phase, a data set was collected from $1/2^{\circ}$/min diffractograms and computer refined using a least squares procedure. The calculated diffraction angles were also useful in deconvoluting overlapping reflections in complex regions of the diffractograms.

Electron microprobe analysis (EMA) was performed with a JEOL JSM-35 scanning electron microscope equipped with a Kevex energy dispersive detector and Tracor Northern analytical/automation data analysis system. A Bence-Albee correction program was used in data reduction for quantitative analyses. Various natural silicate and oxide standards were used. Analyses of grains having similar elemental chemistry and morphology were averaged to insure that the atomic proportions calculated from these analyses would be representative.

RESULTS AND DISCUSSION

Table II gives the chemical analysis of bulk ground gasification ash. This composition is typical of lignite-derived combustion and gasification ashes [8]. In a combustion ash, a loss-on-ignition (LOI) of 5.82% would indicate the presence of considerable unburned coal or char. However, in this gasification ash, LOI is due largely to loss of CO_2 and H_2O by decomposition of calcite and hydroxides/hydrates (portlandite, zeolites and ettringite).

Figure 1 is a portion of the x-ray diffractogram of bulk ground ash. It can be seen by the large number of peaks and by the absence of a prominent diffuse maximum in the region between 20 and 35° that the ash is dominantly crystalline. Twelve of the crystalline phases eventually found in the ash have characteristic peaks in the bulk diffractogram and nine of these are

TABLE II

Chemical Analysis of Lignite Gasification Ash

Oxide	Wt.%	Oxide	Wt.%	Oxide	Wt.%
SiO_2	23.0	Na_2O	8.5	SrO	0.50
Al_2O_3	11.3	K_2O	0.5	MnO	0.16
Fe_2O_3	15.2	TiO_2	0.9	BaO	0.06
CaO	22.7	P_2O_5	0.02	ZrO_2	0.02
MgO	6.8	SO_3	2.25	CuO	0.01
Loss on ignition = 5.82%				Moisture = 0.14%	

indicated on the figure. Table III gives the codes, names and nominal compositions of all phases identified in the METC ash. These have been grouped into four classes that will be discussed later.

Figure 2 is an expansion of the region between 31 and 35° where many important peaks of the refractory phases occur. Through grain-picking, in which grains enriched in only a few of these phases were selected for study by XRD, the presence of these phases was confirmed and minor or trace phases were identified. The deconvolution and assignment of the peaks to particular phases were quite difficult in the complex 32 to 34° region. Deconvolution was aided by indexing selected peaks of many of the phases, calculation of the refined cell parameters and subsequent calculation of the theoretical angles in which peaks known to be intense should occur. Even with all of this effort, not all of the intensity in this region is fully assigned. Several of the minor or trace phases observed only by SEM have strong reflections that occur in this region and may be contributing to the unassigned intensity.

To compare the relative abundances of crystalline phases in the bulk ash and in various fractions of the ash, we prepared a compilation of the intensities of a characteristic peak (usually the strongest) of each phase. This peak intensity compilation for four specimens of bulk ground ash and eleven size fractionated specimens, run at least in duplicate, is given as Table IV. The specimen preparation (powder smeared on a glass slide) and instrument settings (40 KV, 20 ma, 2° speed, 250 cps scale factor and 2 sec. time constant) were kept constant. The height of the peak was read from strip charts. In the one case where the peak height exceeded the chart height, its height was determined at a larger scale factor and normalized to 250 cps.

Before adopting this procedure, we were concerned about the frequent lack of reproducibility in measurements of peak intensities in diffractograms from nominally identical specimen preparations. However, our hypothesis is that in a complex phase assemblage where virtually all grains are mixtures of phases with dimensions typically less than 10 um [5], the effects of preferred orientation and particle statistics on measured intensities would be minor. Except for the peaks of calcite, the heights for the four specimens of bulk ground ash did show good reproducibility for this type of experiment. Calcite has been observed by SEM in relatively large masses [5] in contrast to the small crystallite size and multiphase occurrences of most other ash

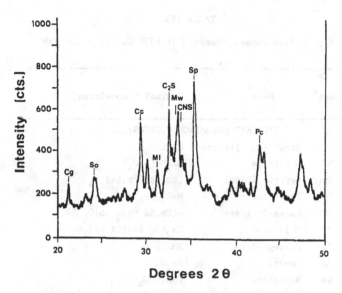

Figure 1. Portion of the X-Ray Diffractogram of METC Gasification Ash

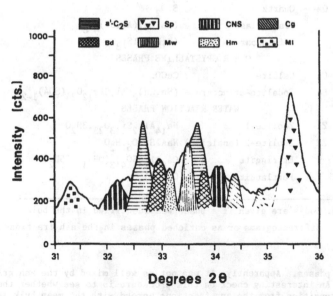

Figure 2. Expanded Portion of the Diffractogram in Figure 1 with Peaks Deconvoluted to Show Individual Phases. (Bd = bredigite, a C_2S phase)

TABLE III

Crystalline Phases Observed In METC Gasification Ash

Code[a]	Name	Nominal Composition
	SILICATES/ALUMINOSILICATES/OXIDES	
C_2S	Dicalcium Silicate	Ca_2SiO_4
Mw	Merwinite	$Ca_3Mg(SiO_4)_2$
Sp	Ferrite Spinel	$(Mg,Fe)(Fe,Al)_2O_4$
Pc	Periclase	MgO
CNS	Ca-Na-Silicate	$\sim(Ca,Na,Mg)_{2.2}SiO_4$
M1	Melilite	$Ca_2(Mg,Al)(Si,Al)_2O_7$
Cg	Carnegieite	$NaAlSiO_4$
Hm	Hematite	Fe_2O_3
Ne	Nepheline	$NaAlSiO_4$
	Mullite	$Al_6Si_2O_{13}$
	Pyroxene	$(Ca,Na)(Mg,Fe)Si_2O_6$
	RESIDUAL COAL MINERALS	
Qz	Quartz	SiO_2
	Plagioclase	$(Na,Ca)(Al,Si)_4O_8$
	K-feldspar	$(K,Na)AlSi_3O_8$
	OTHER CRYSTALLINE PHASES	
Cc	Calcite	$CaCO_3$
So	Sodalite-structure	$(Na,Ca)_8(Al,Si)_{12}O_{24}(SO4)_{1-2}$
	WATER REACTION PHASES	
Z1	Zeolite-1	$Na_{14}Al_{12}Si_{13}O_{23} \cdot 3H_2O$
Z2	Zeolite-2 (analcime)	$NaAlSi_2O_6 \cdot H_2O$
	Ettringite	$Ca_6Al_2(SO_4)_3(OH)_{12} \cdot 25H_2O$
	Portlandite	$Ca(OH)_2$

a. Codes are given if a phase was identified in the bulk
   diffractograms or as enriched phases in the ash fractions.

crystalline phases. Apparently, it was not as well mixed by the ash grinding
procedure. An interesting check on this procedure is to see whether the
weighted intensities from the ash fractions agreed with the mean bulk ash
intensities. For most of the phases, the intensity agreement was within 10-
20% of the mean of the bulk intensities. We consider this agreement
satisfactory for a rapid relative abundance procedure.

TABLE IV

X-Ray Diffraction Results on METC Gasification Ash

ASH SIZE	WT. FRACT.	No.	Sp	C$_2$S	Mw	Pc	CNS	Cc	Ml	So	Cg	Hm	Ne	Qz	Z1	Z2	LOI	MOIST.
BULK ASH		1	68	47	43	33	30	40	21	17	15	17	14	5	8	0		
BULK ASH		2	72	56	43	33	39	33	21	16	17	13	15	8	8	0		
BULK ASH		3	71	61	45	35	39	30	21	17	16	15	14	9	13	0		
BULK ASH		4	80	58	46	42	34	17	24	22	17	18	16	8	14	0		
			--	--	--	--	--	--	--	--	--	--	--	-	--	-		
MEAN OF BULK:			73	56	44	38	36	30	22	18	16	16	15	8	11	0	5.82	0.14
STD. DEVIATION:			4	5	1	4	4	8	1	2	1	2	1	2	3	0		
>9.5 mm	0.112	1	59	58	42	39	36	3	15	12	18	16	5	5	5	0	2.58	0.54
		2	64	53	55	28	47	10	25	13	37	18	6	7	13	0		
		3	37	53	34	49	49	7	17	24	12	16	8	8	10	0		
		4	72	59	62	28	25	12	25	13	36	18	8	8	6	0		
2.8-9.5 mm	0.538	1	72	59	48	30	45	5	13	14	25	16	6	4	4	0	2.36	0.31
		2	75	50	49	32	42	36	8	18	19	16	9	6	9	0		
		3	43	44	40	23	39	11	14	8	10	12	10	6	10	0		
		4	54	47	40	26	37	7	15	9	16	15	10	9	11	0		
2.0-2.8 mm	0.076	1	65	46	49	28	31	16	20	16	12	10	10	9	10	0	3.44	0.42
		2	65	42	42	29	27	24	23	14	11	14	8	7	9	0		
1.0-2.0 mm	0.114	1	67	45	38	30	28	16	30	14	10	14	9	7	9	0	4.39	0.52
		2	56	39	25	34	27	69	19	20	10	14	11	15	10	0		
710-1000 µm	0.035	1	64	39	32	32	27	42	20	17	8	18	9	0	8	0	6.11	1.07
		2	63	36	29	32	27	60	26	23	9	12	10	8	9	0		
500-710 µm	0.024	1	81	38	37	30	28	57	29	20	10	18	8	0	9	7	10.08	1.02
		2	65	32	34	28	27	60	19	23	8	19	10	0	7	6		
250-500 µm	0.040	1	64	32	26	26	29	104	20	18	6	20	16	4	10	12	14.03	1.40
		2	70	29	28	33	26	120	27	12	7	26	15	0	10	10		
125-250 µm	0.022	1	59	24	21	30	21	124	23	17	4	26	9	4	14	15	17.28	1.89
		2	66	21	19	33	20	148	36	18	7	24	13	5	12	17		
63-125 µm	0.018	1	51	18	0	26	23	176	20	20	8	29	12	5	15	19	18.86	2.08
		2	67	24	0	30	21	180	25	16	8	28	12	2	16	14		
45-63 µm	0.008	1	53	17	0	24	17	216	22	22	5	23	13	7	20	25	19.39	2.14
		2	65	16	0	28	21	212	23	23	4	25	14	5	21	27		
<45 µm	0.014	1	64	14	0	28	26	248	18	31	0	28	12	25	26	49	18.83	2.02
		2	66	17	0	26	27	292	17	29	4	26	10	6	28	46		
			--	--	--	--	--	--	--	--	--	--	--	--	--	--	----	----
WEIGHTED MEANS:			61	46	40	30	36	33	17	15	16	18	9	7	9	2	4.70	0.82

CHARACTERISTIC PEAKS (in degrees two-theta for CuK-alpha)

Sp	Ferrite Spinel	35.3	So Sodalite Structure	24.0
C$_2$S	Dicalcium Silicate	32.6	Cg Carnegieite	21.1
Mw	Merwinite	33.4	Hm Hematite	54.1(a)
Pc	Periclase	42.8	Ne Nepheline	29.7
CNS	Ca-Na-Silicate	34.1	Qz Quartz	26.6
Cc	Calcite	29.4	Z1 Zeolite-1	27.6
Ml	Melilite	31.4	Z2 Zeolite-2	27.0

(a) peak read at 54.1 and scaled to height at 33.1 deg.

The data from Table IV lead to the following observations:

1. The ash picks up moisture on handling. The fine fractions especially had much greater $110°C$ weight losses than did the bulk ground ash.

2. Calcite is enriched in the fines. The $800°C$ LOI increases with calcite abundance. Calcite loses CO_2 at about $650°C$ in air. The fines appear to consist largely of loosely bonded larger grains or coatings that were broken off during the mechanical sieving. The zeolites, sodalite and hematite are also enriched in the fines.

3. Merwinite is absent, and C_2S, CNS and Cg are less abundant, in the fines.

4. Ferrite spinel and periclase have about the same relative abundances in all grain sizes.

In excess of 1600 spot or area analyses of the complete range of grain sizes were obtained with the SEM. The SEM has been employed in several essential roles in our ash mineralogy investigations. It provided chemical analyses that led to the determination of structural formulae for the solid solution phases observed by XRD. The primary means of obtaining chemical information on glass phases was SEM analysis. Phases present below the detection limit of XRD are often easily identified with the SEM. The SEM can provide trial compositions for XRD unknowns. The limiting factor in each of these roles is the requirement that crystallite size be generally larger than five micrometers so that the beam is sampling only one phase. Of the phases listed in Table III, several C_2S-like phases, merwinite, many compositions of ferrite spinels, melilite, hematite, quartz, the feldspars and calcite were widespread in their occurrence and had sufficiently large crystals to provide microprobe analyses. Carnegieite, nepheline, sodalite-structure and zeolite phases apparently crystallized in grains too small for single grain analyses. Each also contains major Na-Al-Si, so that they are difficult to distinguish chemically from glass. Several weight percent of periclase is present in the ash and yet we found no individual grains containing only Mg or Mg with minor Ca (some Ca for Mg substitution is suspected from unit cell parameter measurements). Periclase appeared to be intimately mixed with the ferrite spinels. Mullite, pyroxene, ettringite and portlandite were found by XRD in only a few grain picks and were not encountered in the SEM work. Numerous phases were observed with the SEM that were not detectable or could not be unambiguously confirmed by XRD. Among these were two calcium alumino-ferrites. One of these had a composition consistent with the Portland cement clinker phase brownmillerite.

The structural formulae for the five most abundant solid solution phases in the ash are given in Table V. XRD suggested the presence of two C_2S phases, an alpha-prime phase close to the ideal stoichiometry and a bredigite structure phase. EMA studies did not yield any nearly pure C_2S phases. All phases that came close to the A_2BO_4 stoichiometry had substantial Na and Mg substitution. The stoichiometries in Table V indicate that the C_2S phases in the ash have excess Na due to charge balancing substitutions of 2Na for one Ca/Mg. The assignment of the highest Na C_2S-like EMA stoichiometry to the CNS phase is suggested by analogy of several strong, unassigned, peaks to patterns in the Powder Diffraction File. Ferrite spinel solid solution compositions in individual grains varied greatly within the four end member compositions given in Table V. Calcite analyses indicated partial substitution of Sr and Ba for Ca. The average composition was $(Ca_{0.97}Sr_{0.02}Ba_{0.01})CO_3$. Additional microprobe results are given by Stevenson and Larsen [5].

Several of the phases in Table IV would be incompatible under equilibrium conditions. Familiar examples include (quartz + periclase), (quartz +

TABLE V

Electron Microprobe Analyses[a]

MERWINITE

$$(Ca_{2.82}Na_{0.20}K_{0.01}Ba_{0.01})(Mg_{0.94}Fe_{0.11})(Si_{1.97}Al_{0.03})O_8$$
$$3.04 \qquad\qquad 1.05 \qquad\qquad 2.00$$

MELILITE

$$(Ca_{1.75}Na_{0.21}K_{0.01}Fe_{0.03})(Al_{0.65}Mg_{0.27}Fe_{0.12})(Si_{1.43}Al_{0.57})O_7$$
$$2.00 \qquad\qquad 1.04 \qquad\qquad 2.00$$

45 Gehlenite : 35 Akermanite : 20 Na-Melilite ($NaCaAlSi_2O_7$)

DICALCIUM SILICATE

$$(Ca_{1.59}Na_{0.21}Mg_{0.27}Fe_{0.05}Ba_{0.01})(Si_{0.96}Al_{0.03}S_{0.01})O_4$$
$$2.13 \qquad\qquad\qquad 1.00$$

CALCIUM SODIUM SILICATE

$$(Ca_{1.13}Na_{0.59}K_{0.01}Mg_{0.36}Fe_{0.08})(Si_{1.07}Al_{0.03}Ti_{0.01})O_4$$
$$2.17 \qquad\qquad\qquad 1.11$$

FERRITE SPINEL

Magnetite – Magnesioferrite – Spinel – Ulvospinel Solid Solutions

$$Fe_3O_4 \qquad MgFe_2O_4 \qquad MgAl_2O_4 \qquad Fe_2TiO_4$$

---

a. See Stevenson and Larsen [5] for data on the number of analyses and
standard deviation of the analyses for each solid solution phase.

C_2S), and (periclase + melilite). To see how long it would take to form a
steady state assemblage of compatible phases, we fired pellets of bulk ground
ash at 1100°C in air. After 24 hrs. of firing, the diffractogram shows sharp
and more intense peaks with relative abundances quite different from the
original ash. The peak heights (compare to those in Table IV) were: Ml = 124,
Sp = 148, Mw = 70, C_2S = 45, Cg = 39, Pc = 10. The CNS and sodalite structure
phases, quartz, hematite, nepheline and the zeolites were missing and are
presumed to have reacted. The results are of limited value in writing
reactions because of the decomposition of calcite to leave reactive CaO and
because sodium and sulfur oxides have been found (by thermogravimetric
analysis) to volatilize slowly at 1100°C. The alumina and silica released by
decomposition of sodalite and nepheline would be available to react with
merwinite and C_2S to form the increased amounts of melilite. Nevertheless,
one can see the incompatible phases in the process of reacting to form a more
stable assemblage of silicates. It is also likely that some of the increase
in crystallinity is due to devitrification of the glass portion of the ash.

MODEL FOR LIGNITE ASH GENESIS

Based on XRD and SEM characterization of two lignite gasification ashes, the Wellman-Galusha ash described previously [3] and this METC ash, we will propose a model for the genesis of lignite ash in a non-slagging, fixed-bed, gasifier.

Approximately seven percent of the mass of North Dakota lignite remains as ash after gasification. The typical minerals found in Beulah-Zap lignite [9-11] are pyrite, quartz, calcite, feldspars, gypsum and clay minerals (primarily kaolinite with some mica-illite). These minerals supply Fe, S, Al, Si, K, Na, and Ca oxides to the ash. Sodium, calcium and magnesium also occur on ion exchange sites and as chelates in the organic fraction of the lignite. Dolomite, a likely source of Mg, has not been reported. Almost 100% of the Na, 88% Mg, 41% Ca and 57% K appear to be in the organics [11]. Phosphorous and additional sulfur are present also in the organic portion. All of these constituents combine to form the ash. There is a period of 30-45 min. in which the ash components are hot enough for compound-forming reactions (say, above 700°C) and perhaps 10 min. in which local equilibrium on the micrometer scale might be established (in the range 1050-1150°C). Partial melting of this ash at hot spots where the temperature may have exceeded 1250°C is indicated by the hard, dense, centimeter-size clinkers. The cooling ash spends perhaps 10-20 min. in a 100-200 psi steam/O_2/CO_2 atmosphere before it is discharged from the gasifier. In this area, previously formed phases in the ash can be hydrated (zeolites, portlandite), oxidized (hematite) and carbonated (calcite).

Because of the relatively short residence time, some of the more refractory minerals (quartz and feldspars) pass through the gasifier without completely reacting. The infrequent occurrences of mullite and augite-like pyroxene were in vitreous grains where XRD showed a strong diffuse maximum indicative of abundant glass. These glass-rich grains are probably melted quartz-feldspar-clay agglomerates that may have originally been overburden or clay lenses mixed with the lignite during mining.

Iron from the oxidation of pyrite reacts with the readily available Mg and Al to form the homogeneously distributed, but compositionally very diverse, small grains of ferrite spinels. Oxidation of Mg in the lignite also leads to homogeneously distributed small crystals of periclase. The similar relative abundance of ferrite spinel and periclase throughout all of the ash fractions supports these origins.

The sulfur oxides from pyrite oxidation react with Na-aluminosilicates and some CaO to give the sodalite structure phase. The few poor microprobe analyses we obtained suggest that this phase is intermediate to hauyne and nosean in composition. Its cubic cell parameter is somewhat smaller than that of either of these phases. Based on this result and the high CO_2 fugacity in the gasifier, we suspect that the sodalite structure phase may have some carbonate-for-sulfate substitution.

Merwinite, C_2S phases, Ca-Na-silicate and carnegieite form primarily by direct crystallization from the melts in portions of the ash or from devitrification of glass cooled from these melts. The morphology of merwinite especially, as seen by SEM [5], supports this origin. The low abundances or absences of these phases in the fines, which are chiefly the coatings and poorly bonded grains rather than partially melted clinker, are consistent with this hypothesis.

Hematite is primarily a product of oxidation of Fe^{2+}-containing phases. It forms nearer to the surfaces of grains where it is easily removed during the sieving and makes up a relatively greater portion of the ash fines.

Alteration of sodium aluminosilicates and/or glass in the pressurized steam atmosphere forms zeolites near the surfaces of grains. The zeolites become more abundant in the ash fines because of the removal of these coatings.

We suspect that calcite forms from the reaction of lime and CO_2 in the cooling ash. However, if the CO_2 fugacity in the gasifier is high enough, calcite might be a stable phase even at or near the average gasification temperature. Some calcite might also form from surface portlandite carbonation after the ash is discharged from the gasifier.

Based on its small grain size and slight enrichment in the finer fractions, we propose that nepheline is a reaction product of kaolinite and Na_2O.

Melilite, with its solid solution composition between akermanite, gehlenite and Na-melilite, would be a dominant crystalline phase for this bulk composition if the components were homogeneously distributed and reaction time at high temperature were long enough. It would form in part from reaction of the low-silica C_2S and merwinite with high silica quartz and feldspars. The fact that only a few additional hours at 1000-1100°C leads to abundant melilite in the reaction assemblage supports this hypothesis.

This model should be considered as speculative for the present. It will be tested against another ash specimen prepared in the METC gasifier and numerous gasification ashes from the GPGA plant. A recent preliminary XRD examination of the first gasification ash from the GPGA plant indicated the formation of the same assemblage of phases identified in the previously examined Beulah-Zap lignite gasification ashes.

ACKNOWLEDGMENTS

Research funded by Gas Research Institute Contract No. 5082-253-0771 through the North Dakota Mining and Mineral Resources Research Institute. Kalman Pater of the Morgantown Energy Technology Center is thanked for helpful discussions of the METC gasifier. The AAS analyses were performed by D. J. Hassett and XRF analyses were performed by the Energy Research Center of the University of North Dakota. This manuscript benefited from reviews by J. M. Evans and D. Johnson of GRI and editing by R. J. Lauf.

REFERENCES

1. O. Manz, Cem. Concr. Res. 14[4], 513-520 (1984).

2. D. E. Severson, O. E. Manz and M. J. Mitchell, this volume.

3. G. J. McCarthy, L. P. Keller, P. J. Schields, M. P. Elless and K. C. Galbreath, Cem. Concr. Res. 14[4], 479-484 (1984). [Correct Table II on p. 474.]

4. R. J. Stevenson, Cem. Concr. Res. 14[4], 485-490 (1984).

5. R. J. Stevenson and R. A. Larsen, this volume.

6. D. J. Hassett, K. R. Henke, G. J. McCarthy and E. D. Korynta, this volume.

7. K. C. Galbreath and G. J. McCarthy, "Literature Review and Evaluation of the Correlation of Chemical Composition and Vitrification Behavior of Low Rank Coal Ash," in Characterization, Extraction and Reuse of Coal Gasification Solid Wastes, G. H. Groenewold (Ed.), April-June Quarterly Report to the Gas Research Institute, pp. 80-97 (1984).

8. G. J. McCarthy, K. D. Swanson, L. P. Keller and W. C. Blatter, Cem. Concr. Res., 14[4], 471-478 (1984). [Correct Table I on p. 482.]

9. L. E. Paulson, W. Beckering and W. W. Fowkes, Fuel 51, 224-227 (1972).

10. D. Schott, unpublished report, Grand Forks Energy Technology Center, 1980.

11. S. A. Benson, D. K. Rindt, G. G. Montgomery and D. R. Sears, Ind. Eng. Chem. Prod. Res. Dev. 23, 252-256 (1984).

LEGAL NOTICE

CHARACTERIZATION OF A LIGNITE ASH FROM THE METC GASIFIER
II. SCANNING ELECTRON MICROSCOPY

R.J. STEVENSON and R.A. LARSEN
Natural Materials Analytical Laboratory, North Dakota Mining and Mineral
Resource Research Institute, University of North Dakota, Grand Forks, ND
58202

(Received 29 November, 1984; accepted 20 February, 1985; Refereed)

ABSTRACT

Lignite coal ash produced in a stirred, fixed-bed gasifier at the
Morgantown Energy Technology Center (METC) has been studied by scanning
electron microscopy using both morphologic and chemical analyses. Detailed chemical analyses of phases, grains, and traverses across grains
illustrate the heterogeneous nature of the ash. These data support the
concept that most of the ash was formed by partial melting and subsequent
crystallization and agglomeration.

INTRODUCTION

The morphologies and compositions of the microscopic phases that make
up the Morgantown Energy Technology Center (METC) gasification ash are
best observed by scanning electron microscopy (SEM). Although the grain
size of the minerals present is generally less than 10μm, most chemical
analyses of the major minerals present showed only the mineral and no
contribution from the surrounding matrix. In addition to identifying and
determining the chemical composition of the phases present, SEM was also
used to study heterogeneity among the grains of ash as well as the heterogeneity within large individual ash grains.

EXPERIMENTAL

Unground, unsieved splits of the METC gasification ash were embedded
in epoxy. Polished thin sections of these blocks were then prepared; the
final polish was a 1μm alumina powder in a water suspension. The thin
sections were then coated with carbon to prevent charging in the SEM.
Four samples used in x-ray diffraction (XRD) studies were also prepared
for SEM in the same way.

Both area and spot analyses were used to determine the heterogeneity
of the sample and the composition of the phases present. The JEOL 35C was
operated at 15KV accelerating voltage with a 1000pA beam current. During
the 50 second live time collection of the KEVEX EDS spectrum, dead time
was between 25 and 35 percent, depending on the material being analyzed.
A Tracor Northern data analysis system calculated oxide weight percent for
the EDS spectra using a Bence-Albee correction program [1].

Over 270 grains selected for analyses were chosen such that the whole
spectrum of grain sizes [2] would be represented. The size of the grains
that was recorded along with their chemistry was the smallest dimension
observed in the SEM.

The chemical heterogeneity within individual ash particles was determined by traversing each of five large ash grains with a series of 50μm

square beam scan areas. The results of the traverse, therefore, are based on a corridor 50µm wide across the grain. The size of the area is such that effects of large crystallites are not significant since the larger of the phases observed were generally less than 10µm in length and most were in the 5µm size range.

Most of the photography of the samples was done using secondary electron imaging (SEI) at a magnification of 1000x. Lower magnification photographs to provide an overview were also taken, as well as higher magnifications to illustrate and document small features. The photographs also document the location of all of the over 1600 analyses completed on the METC gasification ash.

RESULTS AND DISCUSSION

The SEM characterization of the METC gasification ash can be divided into three studies: inter-grain heterogeneity, intra-grain heterogeneity, and morphological and chemical description of the phases present.

Inter-Grain Heterogeneity

Column 1 of Table I illustrates the overall degree of heterogeneity in the grains of the METC ash. The table lists the mean composition of each sample size fraction and its standard deviation, which is used as a measure of heterogeneity on a inter-grain basis. The degree of heterogeneity in the METC ash is less than that shown for a similar ash (Wellman-Galusha gasification ash [3]).

TABLE I

MEAN WEIGHT PERCENT AND STANDARD DEVIATION FOR METC GASIFICATION ASH

	ALL (275)a	>1000µm (48)	500-1000µm (56)	125-500µm (68)	<125µm (103)
SiO_2	25.96(9.09)b	27.14(7.07)	25.60(7.27)	25.10(7.41)	26.56(11.28)
Al_2O_3	16.31(6.39)	17.29(5.12)	17.42(5.97)	17.00(6.57)	14.97(6.74)
FeO	13.02(11.91)	12.08(3.57)	13.10(8.96)	12.36(7.73)	13.54(16.59)
MgO	6.00(4.02)	6.12(1.74)	6.26(2.56)	6.97(3.55)	5.18(5.36)
CaO	23.75(12.51)	22.00(5.29)	22.29(6.01)	23.41(6.63)	25.81(18.73)
Na_2O	10.82(4.30)	11.23(2.41)	11.35(2.69)	11.04(3.05)	10.04(5.28)
K_2O	0.87(0.60)	0.87(0.59)	0.83(0.52)	0.85(0.52)	0.89(0.69)
SO_3	2.18(2.05)	2.14(0.79)	2.14(1.03)	2.20(1.04)	1.89(1.42)

Notes:
a - number of grains analyzed
b - standard deviation of normalized analyses

The other four columns in Table I illustrate the variable heterogeneity of the METC ash based on the size of grains. These four columns
(Table I) also illustrate the overall homogeneity of the composition of
these size fractions. The three larger (>1000μm, 500 to 1000μm, and 125
to 500μm) size categories show less dispersion than the smallest size
fraction (<125μm). This smaller size fraction is the one that is rich in
calcite [2] and has a standard deviation for CaO three times that of the
other three size fractions. All of the oxides shown in Table I have
larger standard deviations in the <125μm size fraction. SEM observations
indicate that the spinel present in the smaller size fraction is dominantly magnetite (a spinel richer in FeO than that found in the larger
grains), which accounts for the larger increase in FeO dispersion in the
smaller size fraction. The other oxides' increase in standard deviation
in the <125μm size fraction are probably an indirect result of the small
pure phases (calcite and magnetite) present in the smallest size fraction
but also could be a result of heterogeneity of other phases.

The higher number of discreet calcite grains and generally larger
dispersion for <125μm size fraction are also shown in Figure 1. Figure 1
is a ternary plot of $CaO-SiO_2-Al_2O_3$ weight percent and contains the
analyses for the <125μm size fraction as dots and a cross-hatched area in
which the bulk of the >1000μm size fraction is found. The >1000μm size
fraction is typical of all three larger size fractions. All four of the
size fractions tabulated in Table I have the same general trend (Figure
1), although that of the <125μm size fraction is much more pronounced.
This trend is one in which the CaO content of the grains varies independently of SiO_2 and Al_2O_3 and at a relatively fixed SiO_2/Al_2O_3 ratio.
(Analyses that plot near the $SiO_2-Al_2O_3$ edge of Figure 1 are poor in CaO
but relatively rich in Na_2O and SO_3 and are generally of glass composition.)

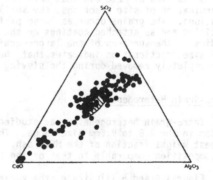

Figure 1
Ternary diagram of weight
percent $CaO-SiO_2-Al_2O_3$ of
analyses of <125μm grains (·)
and the area (striped area)
occupied by the bulk of the
non-char >1000μm grains.

The $Na_2O-Al_2O_3$ weight percent relationship is illustrated in Figure
2. The <125μm size fraction analyses are dispersed over a larger compositional area than the >1000μm size fraction. The trend shows a linkage
of the amount of Na_2O and Al_2O_3 in the grains, that is, the ratio of Na_2O
to Al_2O_3 is constant (approximately 0.73). The >1000μm size fraction has
the same trend, although it is not as well developed.

180

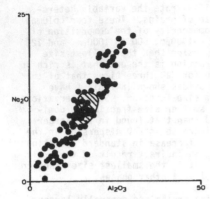

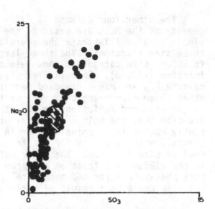

Figure 2
Weight percent Al_2O_3-Na_2O plot
of analyses of <125μm grains (·)
and the area (striped area)
occupied by the bulk of the
non-char >1000μm grains.

Figure 3
Weight percent SO_3-Na_2O
plot of analyses of <125μm
grains (·) and the area
(striped area) occupied by
the bulk of the non-char
>1000μm grains.

There is a positive correlation between Na_2O and SO_3 weight percent
(Figure 3). The ratio is relatively constant for the <125μm size fraction and also for the >1000μm size fraction although less developed
because of the lower dispersion of grain compositions.

On samples of METC ash that have not been mechanically shaken for
determination of size fractions, the smaller size fraction occurs in two
locations. The grains occur as loose particles mixed in with the larger
particles and as attached coatings on the larger grains. Preliminary
studies of the surfaces of the larger grains that have undergone mechanical size fractionation indicates that these coatings still exist and are
not completely removed during the sieving process.

Intra-Grain Heterogeneity

Intra-grain heterogeneity was studied by selecting five grains at
random in the 2.8 to 9.5mm size range. That range represented the
largest weight fraction of the METC ash. All five grains show dispersion
of composition comparable to that of the >1000μm size fraction.

Figures 4 and 5 illustrate the variability of SiO_2, FeO, MgO, Na_2O,
K_2O, and SO_3 weight percent across one of the grains. Different regions
of varying composition are not separated by sharp boundaries as would
occur if smaller grains were fused together just before cooling but
before exchange of elements could occur. Instead, relatively wide,
gradual boundaries exist for this grain and for most grains of this size,
suggesting that these large grains were built up from smaller, heterogeneous grains which had sufficient time, prior to cooling to a solid, to
exchange elements but insufficient time to blend their heterogeneous
compositions completely.

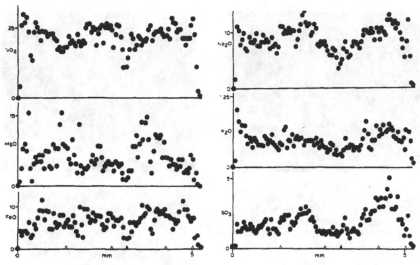

Figure 4
Plots of SiO$_2$-distance, MgO-
distance, and FeO-distance on a
grain traverse. Location of
Figures 6, 7, 8, and 9 shown
as (Λ) from left to right,
respectively. A crack is
present at approximately 3.5mm
(▲).

Figure 5
Plots of Na$_2$O-distance, K$_2$O-
distance and SO$_3$-distance on
the same grain traverse as
in Figure 4.

Correlations between oxide weight percents in the analyses of this grain, that were of the grain itself, and not of the surrounding epoxy or of the coating on the grain, showed generally similar trends to those observed for the whole grains discussed above.

There is an inverse relationship between SiO$_2$ and FeO weight percent in the traverse of the grain. This is due to the variable amount of the dominant FeO phase, ferrite spinel. Where ferrite spinel is a significant constituent of an area there are fewer silicate phases (including glass).

There is a positive correlation between FeO and MgO weight percent. This is because the bulk of the MgO present in the ash is associated with the ferrite spinel phase and not with the silicates.

There is also a positive correlation between Na$_2$O and SO$_3$, Na$_2$O and K$_2$O, and K$_2$O and SO$_3$. These relationships are such that the relative amounts of these oxides in analyses along the traverse are nearly constant. This is shown in the traces of Na$_2$O, K$_2$O, and SO$_3$ (Figures 5). The dominant phase containing all of these oxides is the interstitial glass. The traces of Na$_2$O, K$_2$O, and SO$_3$ or FeO and MgO can be used to estimate the relative abundance of glass and ferrite spinel, respectively, in the traversed grains.

Figure 6
SEI of area at 1.40mm on grain
traverse. Brownmillerite (Bm)
and spinel (Sp) in a glass-like
matrix. Scale bar is 10.0μm
long.

Figure 7
SEI of area at 2.70mm on
grain traverse. Melilite
(Ml), spinel (Sp) and glass
(G) are the major phases.
Scale bar is 10.0μm long.

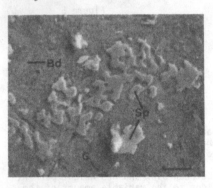

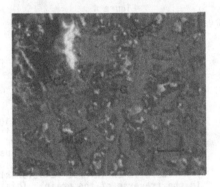

Figure 8
SEI of area at 4.25mm on grain
traverse. Bredigite (Bd),
spinel (Sp) and glass (G) are
the major phases. Scale bar
is 10.0μm long.

Figure 9
SEI of area at 5.02mm on
grain traverse. Merwinite
(Mw), spinel (Sp), glass
(G), and calcite (Cc) are
the major phases. Scale
bar is 10.0μm long.

Photomicrographs of four areas in the traversed grain (Figures 6, 7,
8, and 9) illustrate the heterogeneity of the larger grains with respect
to morphology as well as the amount of glass, ferrite spinel, and other
phases present. In Figures 6, 7, 8, and 9 the most obvious difference
between the areas is the morphology of the ferrite spinel, ranging from
euhedral to anhedral. The other phases present also vary in morphology
and size. The major phases present in the traversed grain are ferrite
spinel, melilite, glass, merwinite, bredigite (a C_2S phase), probable
brownmillerite, CNS (calcium sodium silicate [2]), and calcite in the

coating material. A further discussion of morphology and chemistry of the minerals present follows in the next section. The locations of the photographs are indicated in Figures 4 and 5.

Phase Chemistry

Table II contains the average compositions and standard deviations of the major phases of the METC ash. Ferrite spinel and glass are not shown because of their variability. In some analyses ferrite spinel has excess MgO for magnesioferrite ($MgFe_2O_4$). This excess MgO is thought to be present as periclase [2]. The melilite (Ml) average analysis can be described by the akermanite-gehlenite-sodium melilite end members of the melilite solid solution series [2].

Merwinite (Mw), CNS, and bredigite (Bd) are structurally related [2] and chemically similar. The chemical difference between the three minerals can be illustrated on a MgO-CaO-Na$_2$O weight percent ternary diagram (Figure 10). The bredigite phase plots closer to a pure C$_2$S chemistry (at the CaO apex) than either of the other two phases, and CNS is distinguished from the other two by its relatively large amount of Na$_2$O. Between the merwinite and bredigite areas there were a few analyses that seem to be intermediate in composition between these two phases. These analyses may represent a different phase but more likely they represent a mixture of merwinite and bredigite on a small scale.

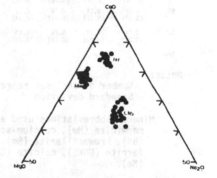

Figure 10
Plot of top half of weight percent MgO-CaO-Na$_2$O ternary diagram showing the compositional space occupied by merwinite (Mw), bredigite (Bd), and CNS.

The other minerals tabulated in Table II are probable brownmillerite (Bm) and unknown 5 (UNK5) both calcium alumino-ferrites, calcite with 0.79 weight percent BaO and 1.67 weight percent SrO, and sodalite (So) and nepheline (Ne). The latter two analyses are tabulated to illustrate the small size of these phases in the ash. The mineral size is so small that these analyses are the best obtained and yet they contain over 6 weight percent FeO and MgO, which exceeds the solubility in these phases.

Figures 11, 12, 13, and 14 along with Figures 6, 7, 8, and 9 combine to illustrate the heterogeneous texture and morphology of the METC ash and its minerals. Figure 11 is of a small (<125μm) grain composed of calcite and glass. Also note the small euhedral calcite grain to the left side of the photograph.

TABLE II

MINERALOGY OF METC GASIFICATION ASH

	Ml (20)[a]	Mw (64)	CNS (26)	Bd (38)	Bm (6)	UNK5 (5)	Cc (13)	So (3)	Ne (1)
SiO_2	31.08 (3.38)[b]	35.50 (1.14)	36.92 (1.63)	32.23 (0.84)	7.49 (1.05)	1.03 (0.39)	0.25 (0.22)	31.90 (3.13)	34.03
Al_2O_3	22.49 (4.39)	0.42 (0.48)	0.71 (0.49)	0.82 (0.55)	24.08 (1.67)	16.44 (0.69)	0.20 (0.16)	28.21 (3.52)	28.03
FeO	3.90 (1.42)	2.29 (0.82)	3.43 (0.84)	1.96 (0.44)	4.80 (1.27)	31.45 (1.24)	0.41 (0.21)	1.72 (1.07)	3.22
MgO	3.91 (1.40)	11.41 (0.48)	8.59 (1.34)	5.55 (0.48)	0.56 (0.44)	1.57 (1.77)	0.06 (0.10)	1.06 (1.95)	2.78
CaO	35.66 (2.57)	47.39 (1.65)	37.50 (1.49)	50.36 (1.47)	35.59 (1.86)	44.66 (1.04)	54.83 (2.11)	3.45 (1.88)	2.15
Na_2O	2.36 (0.94)	1.85 (0.54)	10.87 (0.84)	3.70 (0.66)	1.44 (0.71)	0.64 (0.54)	0.47 (0.26)	14.46 (0.85)	10.39
K_2O	0.12 (0.10)	0.14 (0.10)	0.23 (0.07)	0.24 (0.09)	0.11 (0.07)	0.06 (0.06)	0.0 (0.0)	0.93 (0.17)	0.47
TiO_2	0.10 (0.15)	0.13 (0.18)	0.29 (0.26)	0.13 (0.23)	0.11 (0.13)	1.59 (0.98)	0.0 (0.0)	0.14 (0.17)	0.22
P_2O_5	0.0 (0.0)	0.03 (0.08)	0.01 (0.04)	0.01 (0.04)	0.05 (0.12)	0.04 (0.08)	0.01 (0.04)	0.0 (0.0)	0.0
MnO	0.01 (0.03)	0.04 (0.08)	0.05 (0.09)	0.04 (0.07)	0.07 (0.08)	0.69 (0.06)	0.07 (0.11)	0.0 (0.0)	0.0
SO_3	0.22 (0.30)	0.14 (0.22)	0.10 (0.14)	0.31 (0.34)	0.92 (0.54)	0.0 (0.0)	0.17 (0.21)	5.97 (0.85)	1.10
BaO	0.24 (0.21)	0.33 (0.22)	0.28 (0.28)	0.65 (0.28)	0.16 (0.17)	0.16 (0.22)	0.79 (0.61)	0.0 (0.0)	0.0
SrO	--	--	--	--	--	--	1.67 (1.22)	--	--

Notes:
  a - Number of analyses represented
  b - Standard deviation

  Mineral abbreviations used are as follows: melilite (Ml), merwinite (Mw), calcium-sodium silicate (CNS), bredigite (Bd), brownmillerite (Bm), unidentified calcium-alumina ferrite (UNK5), calcite (Cc), sodalite (So), and nepheline (Ne).

Figure 12 is composed of spinel and glass. The dendritic morphology of the spinel illustrates the once molten nature of some of the grains in the ash.

Figure 13 is composed of merwinite, bredigite, CNS, glass, and spinel. The merwinite forms the cores of the elongated crystals while bredigite and CNS are seen to mantle the merwinite cores. The glass present is Na_2O-rich (>20 weight percent of the glass is Na_2O).

Figure 14 shows another morphology of merwinite. In this case the crystals of the merwinite are hollow.

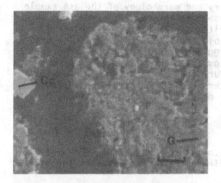

Figure 11
SEI of METC ash showing habit
calcite (Cc) and glass (G).
Scale bar is 10.0μm long.

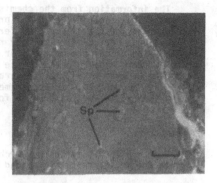

Figure 12
SEI of METC ash showing
spinel (Sp) in a grain of
glass. Scale bar is 10.0μm
long.

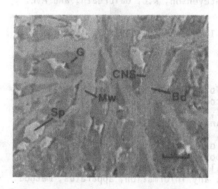

Figure 13
SEI of METC ash showing habit
of merwinite (Mw), bredigite
(Bd), CNS, spinel (Sp), and
glass (G). Scale bar is
10.0μm long.

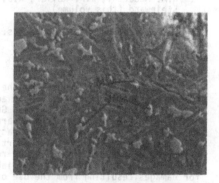

Figure 14
SEI of METC ash showing
habit of spinel (Sp) and
hollow merwinite (Mw).
Scale bar is 10.0μm long.

Figure 6 illustrates an occurrence of brownmillerite while Figure 9
shows another type of morphology of merwinite. The merwinite morphology
shown in Figure 9 is more common than that shown in Figures 13 and 14.
Also of note in Figure 9 is the location of the calcite which occurs
along the edge of the traversed grain as a late stage coating.

CONCLUSIONS

This sample of a lignite gasification ash shows heterogeneity on
three levels. The first two levels are the heterogeneity with respect to
chemistry at inter- and intra-grain levels. The third is the heterogeneity of the morphology of the minerals including hollow, euhedral, or
mantled merwinite and ferrite spinel that ranges from euhedral to dendritic.

The information from the chemistry and morphology of the ash sample indicates that some of the grains present in the sample formed from a liquid or at least a liquid plus ferrite spinel crystal assemblage, followed by a period of relatively rapid cooling. Other particles are small grains of calcite or glass or agglomerations thereof. The larger grains appear to have formed by the agglomeration of smaller grains and, depending on residence time, temperature, or other factors, become partially homogenized. These parameters could be controlled to produce an ash that would be tailor-made for disposal and/or utilization.

ACKNOWLEDGEMENTS

Research funded by Gas Research Institute Contract No. 5082-253-0771 through the North Dakota Mining and Mineral Resources Research Institute.

REFERENCES

1. A.E. Bence and A. Albee, J. Geol., 76, 382-403 (1968).

2. G.M. McCarthy, L.P. Keller, R.J. Stevenson, K.C. Galbreath, and A.L. Steinwand (this volume).

3. R.J. Stevenson, Cem. Concr. Res., 14:4, 485-490 (1984).

LEGAL NOTICE

CHARACTERIZATION OF A LIGNITE ASH FROM THE METC GASIFIER
III. CORRELATIONS OF LEACHING BEHAVIOR AND MINERALOGY

D.J. HASSETT*, G.J. McCARTHY**, K.R. HENKE***, and E.D. KORYNTA****
  * Fuels Analysis Laboratory, University of North Dakota, Grand Forks,
    ND 58202
  ** Department of Chemistry, North Dakota State University, Fargo, ND
    58105
  *** Energy Research Center, University of North Dakota, Grand Forks, ND
    58202
  **** Engineering Experiment Station, University of North Dakota, Grand
    Forks, ND 58202

(Received 29 November, 1984; accepted 6 February, 1985; Refereed)

ABSTRACT

Lignite gasification ash from the Morgantown Energy Technology
Center (METC) gasifier was subjected to two short-term leaching
treatments. The concentrations of regulated elements in the EPA EP
procedure leachate from the bulk METC ash did not exceed the "EP Trigger
Limits." A modification of this procedure that uses a basic synthetic
groundwater instead of the acid EP extractant was also performed on the
bulk ash and its eleven size fractions. Water equilibria modeling was
used to explain the concentrations of major elements in solution.
Concentrations of minor and trace elements in solution after leaching
with synthetic groundwater were also below "EP Trigger Limits."

INTRODUCTION

It is widely appreciated that coal gasification ash contains small
amounts of regulated elements (As, Ba, Cr, Se etc.). In addition, some
of its major species (Na, Ca, SO_4) are present in soluble or slightly
soluble phases that can lead to degradation of groundwater quality
through increases in total dissolved solids. For these reasons,
considerable care is used in the disposal of gasification ash. For
example, clay-lined pits located at the adjacent lignite mine are used at
the Great Plains Gasification Associates site in Beulah, North Dakota.
The primary objective of our research program is to lower or eliminate
the cost of disposal by finding economical uses for some or all of the
gasification ash [1,2]. We expect that the ash utilization options under
study [2] could substantially lower the possibility of harmful releases,
through incorporation of the ash into new solid forms having one or more
of the following characteristics: lower specific surface areas, lower
concentrations of the elements of concern, lower leachabilities for these
elements because they are now incorporated in more stable host phases.
Other options (e.g. aggregate for road bases) might use only particular
size fractions of the ash. This would leave a reduced mass of ash for
disposal. Finally, it is possible that the larger size fractions
consisting of dense hard grains rich in silicates, oxides, and
aluminosilicates might have such low leachabilities of the regulated
elements that these fractions could be removed for disposal by less
expensive means, for example, at Beulah by burial in the overburden
during reclamation of the strip mine. Data given in Part I of this
report [3] show that 65% of the mass of METC ash consisted of clinkers or
agglomerates larger than 2.8 mm, a size that might be separable by
screening.

We have included leaching studies in our ash characterization work for two primary reasons: first, to establish a baseline for comparison of the leaching behavior of gasification ash with that of products made from it, and second, to determine the relative leaching potential of the ash size fractions. The behavior of the ash on contact with various aqueous solutions will also add insight into mechanisms of ash reactions during fabrication or processing of some potential products (e.g. pozzolanic or cementitious reactions). For these reasons, we are especially interested in correlating leaching behavior with the knowledge of ash mineralogy and morphology gained from x-ray diffraction and electron microscopy. The long term objective is to make the results on lignite gasification ash leaching behavior as generic as possible by determining the actual crystalline or noncrystalline phases that control the release of the elements of concern. Our first results on the METC ash presented here were obtained with a relatively short-duration regulatory test (the EP batch leaching method) and with a variation on this test that uses a basic leaching solution (an artificial groundwater) that is more representative of ash disposal in the Nothern Great Plains.

EXPERIMENTAL PROCEDURES

McCarthy et al. [3] gave a description of the specimen handling and size separations. Portions of the same specimens studied by XRD were used in the leaching tests. One additional set of tests was performed on a batch of the size fraction between 2.8 and 9.5 mm freshly ground to -200 mesh. This was done to expose fresh surfaces and set a possible upper limit on leachability of the hard and dense grains of the large size fraction.

The leaching test adopted for this study was the proposed U.S. EPA Extraction Procedure (EP) method [4]. To a 100 g resentative specimen of the METC ash, 1.6 L of deionized water was added. The procedure calls for the pH to be adjusted to 5.0±.2 with 0.5 N acetic acid up to a limit of 4 mL of acid per gram of solid. The full 400 mL called for in the test was added at intervals, but the final filtered pH had only dropped to 5.6. The mixture was agitated on rotary mixers for 24 hours at approximately 25°C with the exclusion of air. At the end of the extraction period, deionized water was added to bring the total volume of the liquid to 2 L. The mixture was filtered through a fine glass fiber prefilter and a 0.45 μm filter under 50 psi of argon. The filtered leachate was preserved with nitric acid and refrigerated until analyzed. Elemental concentrations of the leachant were determined by atomic absorption spectrometry (AAS) or inductively coupled argon plasma spectrometry (ICAP).

The EPA EP method is an acid leaching test that has little relevance to the ability of groundwater in the Upper Great Plains to extract elements from lignite gasification ash. Such an ash will quickly establish a pH between 10 and 12.5 when mixed with deionized water. Groundwater in the region typically has a pH greater than 7.5. For these reasons, the EP procedure was modified to use a synthetic groundwater (SGW) instead of an acetic acid solution. This SGW was prepared with 1.00 g of Na_2SO_4 and 2.0 g of Na_2CO_3 per 2 L of deionized water. The SGW to solid ratio, extraction time, filtering and analysis procedures were the same as those of the EP method. The eleven size fractions as well as the bulk ash were tested with this synthetic groundwater. It was necessary to use smaller masses of sample for the less abundant ash fractions, but the correct solution to solid ratio was maintained. Most analyses were obtained by ICAP less than 24 hours after filtering.

RESULTS AND DISCUSSION

Table 1 gives the results of one EPA EP batch leaching test on a representative specimen of METC gasification ash. Results obtained from the modified procedure using SGW are given in Table 2.

TABLE 1

EP BATCH LEACHING RESULTS FOR THE BULK METC ASH
(PARTICLES >9.5mm WERE NOT CRUSHED)

mL of 0.5N acetic acid added:				400mL
Filtered pH:				5.6
g of ash:				100g
volume of extracting solution:				2 L
Elements (mg/L):	Si	18	Cd	<0.1
	Al	15	Cr	<0.03
	Fe	3.48	Cu	<0.02
	Ca	1400	Mn	1.00
	Mg	129	Mo	0.032
	Na	586	Ni	<0.05
	K	30	Pb	<0.1
	Ag	NA	Se	<0.002
	As	0.031	Sr	35
	B	12	Zn	<0.2
	Ba	1.86	Hg	<0.0004
	Be	NA		

NA = Not Analyzed

For comparison, we have collected four sets of standards to provide a context in which to evaluate the leaching behavior of the METC ash. Table 3 contains three water quality standards, National Interim Primary Drinking Water Standards [5], National Secondary Drinking Water Regulations [6], Long Term Irrigation Standards [8]. Also listed are the EP "trigger concentrations" [4] that apply to seven of the elements in Tables 1 and 2. The concentrations of the elements extracted from the ash with the EP procedure are well below the "EP Trigger Values" to which this procedure applies.

In the remainder of this section we will consider the individual elements extracted by the SGW procedure and note whether they exceed any of the standards listed in Table 3. In the case of the major elements, we will suggest what phases in the ash are controlling the releases of these elements based on conventional water equilibrium modeling and geochemical considerations. Note that the situations in which the drinking water or irrigation standards would apply (a water supply well sunk into an ash pit saturated with groundwater) are not especially realistic and are discussed here only for baseline comparison purposes.

TABLE II

SYNTHETIC GROUNDWATER LEACHING RESULTS (mg/L) FOR THE METC BULK ASH AND SIZE FRACTIONS

Sample	Bulk	>9.52mm	2.8mm	2.8mm Ground	Sieve Size 2.0mm	1.0mm	710µ	500µ	250µ	125µ	63µ	45µ	<45µ	Groundwater Blanks – Quality Control Grdwater #1	Grdwater #2	Grdwater #4	Grdwater #5	Grdwater #6
Filtered pH	11.1	10.0	10.2	12.1	11.3	11.4	11.3	11.2	11.1	11.2	10.8	11.3	11.2	8.4	8.4	8.6	8.5	8.5
g of ash	100	50	100	100	30	100	40	40	40	40	30	10	30	0	0	0	0	0
Grd water Batch #	5	5	4	6	5	5	1	1	1	1	2	5	1	1	2	4	5	6
Si	43	9	19	65	68	61	49	43	35	37	31	69	38	<0.1	<0.1	<0.1	<0.1	<0.1
Al	0.9	2.7	0.3	50.3	1.1	0.7	0.6	0.4	0.5	0.5	0.4	0.6	1.1	<0.1	<0.1	0.1	0.1	<0.1
Fe	0.54	0.08	0.02	0.02	0.08	0.22	0.29	0.24	0.34	0.43	NA	0.50	1.26	<0.01	<0.01	<0.01	<0.01	<0.01
Ca	2.8	2.4	2.4	5.0	2.2	2.7	3.0	2.5	2.7	3.4	5.0	4.1	5.3	<0.1	<0.1	<0.1	<0.1	<0.1
Mg	0.46	0.66	1.29	NA	0.22	0.16	0.24	0.20	0.26	0.36	0.37	0.43	0.89	0.01	0.01	<0.01	<0.01	<0.01
Na	853	543	605	928	665	733	907	954	1081	1234	1264	1373	1300	430	428	412	406	416
K	24	10	9	40	16	22	25	25	30	30	30	53	30	<0.1	<0.1	0.3	<0.1	<0.1
As	0.172	0.039	0.062	NA	0.118	0.136	0.182	0.207	0.254	0.383	0.352	0.455	0.358	<0.001	<0.001	<0.001	<0.001	NA
B	6.78	1.57	2.65	7.85	5.72	6.57	6.59	6.65	7.17	9.21	9.45	12.31	8.79	<0.05	<0.05	<0.05	<0.05	<0.05
Ba	0.105	0.126	0.082	0.067	0.196	0.087	0.179	0.169	0.185	0.197	0.187	0.494	0.289	<0.005	<0.005	<0.005	<0.005	<0.005
Be	NA	NA	<0.004	NA	NA	NA	0.004	<0.004	<0.004	<0.004	<0.004	NA	<0.004	<0.004	<0.004	<0.004	<0.004	NA
Cd	<0.1	<0.1	<0.1	<0.1	<0.1	<0.1	<0.03	<0.1	<0.03	<0.03	<0.03	<0.1	<0.1	<0.1	<0.1	<0.1	<0.1	<0.1
Cr	<0.03	<0.03	<0.03	<0.03	<0.03	<0.03	<0.03	<0.03	<0.03	<0.03	<0.03	<0.03	<0.03	<0.03	<0.02	<0.03	<0.03	<0.03
Cu	<0.02	0.09	0.02	<0.02	<0.02	0.03	0.04	0.04	0.04	0.05	0.07	0.18	0.19	<0.02	<0.02	0.02	<0.02	<0.02
Mn	<0.02	<0.02	<0.02	0.02	<0.02	0.02	<0.02	<0.02	<0.02	0.02	<0.02	0.02	0.04	<0.02	<0.02	0.02	<0.02	<0.02
Mo	0.083	0.017	0.024	NA	0.054	0.068	0.086	0.095	0.117	0.146	0.193	0.191	0.189	<0.01	<0.01	<0.01	<0.01	NA
Ni	<0.05	<0.05	<0.05	<0.05	<0.05	<0.05	<0.05	<0.05	<0.05	<0.05	<0.05	<0.05	<0.05	<0.05	<0.05	<0.05	<0.05	<0.05
Pb	<0.1	<0.1	<0.1	<0.1	<0.1	<0.1	<0.1	<0.1	<0.1	<0.1	<0.1	<0.1	<0.05	<0.1	<0.1	<0.1	<0.1	<0.1
Se	<0.002	<0.002	NA	NA	<0.002	0.0026	<0.002	0.0022	0.0037	<0.002	<0.002	<0.002	<0.002	<0.002	<0.002	<0.002	<0.002	NA
Sr	0.19	0.23	0.25	0.15	0.22	0.23	0.20	0.17	0.16	0.19	0.23	0.25	0.77	<0.2	<0.2	<0.2	<0.2	<0.2
Zn	<0.2	<0.2	<0.2	<0.2	<0.2	<0.2	<0.2	<0.2	<0.2	<0.2	<0.2	<0.2	<0.2	<0.2	<0.2	<0.2	<0.2	<0.2
Hg	<0.0003	NA	<0.0003	NA	NA	<0.0003	NA	NA	NA	<0.0003	NA	NA	NA	NA	NA	NA	NA	NA

NA = Not Analyzed

TABLE 3

EPA EP "TRIGGER CONCENTRATION" AND WATER QUALITY STANDARDS
(All values in mg/L)

Element	EPA Hazardous Waste No.	EP Trigger Concentration RCRA Standards [4]	Primary Drinking Water [5]	Secondary Drinking Water [6]	Long Term Irrigation [8]
Dissolved Solids				500	5000
Sulfate				250	
Aluminum					5.0
Arsenic	D004	5.0	0.05		1.0
Barium	D005	100			
Boron					0.75
Cadmium	D006	1.0			0.01
Chromium	D007	5.0	0.05		0.01
Copper				1.0	0.2
Iron				0.3	5.0
Lead	D008	5.0	0.05		5.0
Manganese				.05	0.2
Mercury	D009	0.2	.002		
Molybdenum					.01
Nickel					.2
Selenium	D010	1.0	.01		
Silver	D011	5.0	.05		.10
Vanadium					
Zinc				5.0	2.0

Silicon

Si in solution varied from 9 mg/L in the largest ash grains to 69 mg/L in the grains retained on the 45 μm sieve. This element is present in most of the phases in METC ash. Kinetically, it is probably the glass phases that control Si solubility in the early stages of leaching before the nesosilicates (C_2S, merwinite), sorosilicates (melilite) and tectosilicates (nepheline, carnegieite, sodalite) can begin to show appreciable solubility. The glass content of the larger grains is probably highest as these are the hard, dense clinkers formed by partial melting of the ash. These grains also leach to give the highest dissolved Si concentrations. The very largest grains give lower Si concentrations, probably because of a much lower specific surface area. The solutions are well undersaturated with respect to amorphous silica as the Si solubility controlling species because above a pH of 10, the concentration of Si in solution can be much higher [7] than the values observed in these leachates. The 118 mg/L Si concentration of the EP leachates is a factor of two higher than the 58 mg/L [7] predicted by equilibrium with amorphous silica at pH < 10.

Aluminum

Al is present in the glass phases, ferrite spinel, nepheline, carnegieite, melilite, sodalite-phase and the zeolites. These are all relatively low solubility phases. The solubility of Al is more likely to be controlled by its oxyhydroxides and hydroxide species and is thus pH dependent. Because of its amphoteric behavior, Al solubility increases at high pH [7]. The freshly ground +2.8 mm fraction gave a leachate with the highest pH recorded in the study and it also had the highest Al concentrations. Al solubility also increases at lower pH. Thus, the Al concentrations are higher in the acid extraction EP solutions.

The Al concentration of all of the SGW procedure solutions except the freshly ground +2.8 mm fraction fall well below the long term irrigation standard.

Iron

As with Si and Al, Fe occurs in many phases. Any breakdown of Fe-containing phases would still leave the Fe solubility controlled by the oxyhydroxides and hydroxides of Fe. Under oxidizing conditions, such as those in our leaching studies, and pH between 10 and 12, Fe solubilities should be even lower than those of Al [7]. The less than 1 mg/L values for Fe in Table 2 are consistent with this mechanism. Note that Fe solubility would be higher if solutions (e.g. groundwater) contacting the ash were reducing. Again, as with Al, the lower pH in the EP test resulted in higher Fe concentrations in solution.

The iron concentration of the bulk ash SGW leachate and the leachate from the smaller ash fractions slightly exceed the National Secondary Drinking Water limit.

Calcium

The concentration of Ca is relatively constant among the bulk ash and the ash fractions. It increases in the freshly ground +2.8 mm fraction and the finest fractions. Ca is a constituent of numerous phases, but the phase that appears to be controlling Ca concentration in the solutions is calcite. It is present in the bulk ash at the 5-10 wt.% level and becomes more abundant in the fine fractions. The solubility of calcite is strongly pH dependent. Routine equilibrium calculations [7] lead to a Ca concentration of 4.8 mg/L at pH = 10 and 2.8 mg/L at pH = 12 from the solubility of calcite in pure water at 25°C in contact with air. These values would be somewhat higher in the moderate ionic strength leaching solutions of this study. The larger size ash fractions are undersaturated with respect to calcite but the smaller fractions, where calcite is more abundant and the available surface areas are greater, give solutions approaching calcite saturation. The solubility of calcite at pH = 5.6, the value for the EP leachant, would give about 700 mg/L of Ca in solution. This explains the high Ca concentration (1400 mg/L) of that solution.

Magnesium

Magnesium concentrations were generally less than 1 mg/L in the SGW leachates. Merwinite, ferrite spinel, periclase, the C_2S phases, melilite and the glass phases all contain magnesium. Among the crystalline phases, periclase is the most soluble. Mg from periclase or glass dissolution would be controlled by the solubility of $Mg(OH)_2$ which, again, is pH dependent. The solubility values for $Mg(OH)_2$ in pure water are 37 mg/L at pH = 10, 0.37 mg/L at 11 and 0.004 mg/L at 12. The SGW leachates would appear to be nearly saturated in Mg with respect to

$Mg(OH)_2$. At pH = 5.6, $Mg(OH)_2$ is very soluble so the much higher Mg values (129 mg/L) found in the EP leachate are reasonable.

Sodium and Potassium

The alkali elements Na and K occur in nepheline, carnegieite, Ca-Na-silicates, sodalite-phase, plagioclase, zeolites and the glass phases in the METC ash. The zeolites detected by McCarthy et al. [3] in the finest ash fractions are both Na-aluminosilicates and would be expected to ion-exchange Na and K readily upon contact with water containing divalent cations. Accordingly, the concentrations of these alkalis do increase significantly with decreasing grain size in the ash fractions. The principal mechanism for the release of Na and K from the aluminosilicates is protonation/ion exchange as the first step in the dissolution of the aluminosilicate framework of the structure. The general stability of these phases is lower at the high pH (10-12) of the SGW solutions than at the mildly acid conditions of the FP solutions. This could explain why the Na concentrations are about 50% greater in the SGW leachates than in the EP leachate.

Na and K make up the bulk of the cation portion of total dissolved solids (TDS). Assuming that sulfate (not measured in the leachates) is the counter ion, the TDS is about 1500 mg/L in the SGW leachates. This is considerably above the Secondary Drinking Water limits but well below the Long-Term Irrigation limits.

Other Elements

Beryllium, cadmium, chromium, nickel, selenium, mercury, zinc and lead are below or near analytical detection limits in the METC leachates. Many elements, such as Be, are not abundant in coal or coal wastes. Mercury and selenium may be abundant in coal, but are too volatile to remain with the gasifier ash. Barium is present in trace amounts in all of the leachates. $BaSO_4$, barium silicates and other common barium phases are very insoluble in water at pH values of 5-12.

Most leachates contained copper at concentrations less than 0.05 mg/L. The SGW leachates from fine ash fractions contained 0.05 - 0.19 mg/L copper. The iron concentrations in these leachates were also usually high. Magnetic enhancement studies of federal samples at University of North Dakota Energy Research Center (UNDERC) indicated that copper is more common in spinel than other transition metals (Cr, Zn, Mn, and Ni). Traces of copper also commonly occur in pyrite. During combustion, the copper-bearing pyrite was probably transformed into copper-bearing spinel. The leaching of copper-iron spinel may have been responsible for the trace concentrations of these metals in the synthetic groundwater leachates.

RCRA Standards and Water Quality Criteria

None of the METC ash leachates (Table 2) violate the RCRA standards for As, Ba, Cd, Cr, Pb, Hg, or Se (Table 3). Silver concentrations have not yet been determined, but are not expected to exceed the standard. Barium, and perhaps other elements, exceed water EP quality criteria in the preliminary EP leachate (Table 3). The EP quality criteria are 0.01 of the RCRA standards.

194

All of the leachates exceeded the irrigation criteria for boron (0.75 mg/L). Aluminum, iron, copper molybdenum, and manganese irrigation criteria were exceeded in at least one EP leachate (Table 1). At least one of the synthetic groundwater leachates exceeded the irrigation criteria for Al, Cu, Fe, and Mo (Table 2).

ACKNOWLEDGMENTS

Research funded by Gas Research Institute Contract No. 5082-253-0771 through the North Dakota Mining and Mineral Resources Research Institute.

REFERENCES

1. O. Manz, Cem. Concr. Res. 14 [4], 513-520 (1984)
2. M. J. Mitchell, O. E. Manz and D. E. Severson (this volume)
3. G. J. McCarthy, L. P. Keller, R. J. Stevenson, K. C. Galbreath and A. L. Steinwand (this volume)
4. U.S. Environmental Protection Agency, Federal Register, No. 45 (98) 33063-33285 (1980)
5. 40 CFR Pt. 141
6. 40 CFR Pt. 143
7. K. B. Krauskopf, Introduction to Geochemistry, 2nd Ed., McGraw-Hill Book Company, New York (1979)
8. W.A. Duval et al. FGD Sludge Disposal Manual, EPRI FP977, RP786-1, January, 1979

COMPARATIVE ECONOMICS OF SEVERAL ALTERNATIVES
FOR BULK UTILIZATION OF FLY ASH AND COAL GASIFICATION ASH

D.E. SEVERSON*, O.E. MANZ**, M.J. MITCHELL***
* University of North Dakota, Department of Chemical Engineering, Grand Forks, North Dakota 58202
** University of North Dakota, Department of Civil Engineering, Grand Forks, North Dakota 58202
*** University of North Dakota, Energy Research Center, Box 8213, University Station, Grand Forks, North Dakota 58202

(Received 29 November, 1984; accepted 25 February, 1985; Refereed)

ABSTRACT

Fly ash and gasification ash have been evaluated economically for the following uses: partial replacement of Portland cement; mineral wool; blended cement; Sulfurcrete®, high flexural strength ceramic products; ash to upgrade roads; glazed ceramic wall tile; and unglazed floor tile. The ash evaluated is a high-calcium, high-sodium material derived from the Beulah-Zap lignite mined in the Beulah region of North Dakota. Of the uses examined, concrete replacement provided an 8.0% cost saving, blended cement 37.3%, high flexural strength ceramics 52.8%, ash in road construction 44.4%, and wall tile 5.2%. Mineral wool had no replacement savings calculated because blast furnace slag is not available locally to provide a consistent basis. Sulfurcrete® did not provide a cost saving over concrete but its use life and properties are sufficiently different from those of concrete to justify use in some applications, provided that the raw materials are readily available.

INTRODUCTION

A paper by Oscar Manz [1] presented background material and proposed that Western fly ash be used in: 1) mineral wool, 2) blended hydraulic cement, 3) high flexural strength ceramic products, 4) ceramic glazed wall tile, and 5) dry pressed ash brick. As a continuation of this project, the following have been added: 6) ash mixed concrete (cement replacement), 7) Sulfurcrete® and 8) ash in haulage or county level road upgrades. Recently we have concentrated on materials that could include gasification ash and concluded a preliminary economic study on selected products in each area.

In the present report, a number of possible alternative uses for fly ash and gasifier ash to be generated by existing and potential synfuels plants are examined from the standpoint of cost feasibility for use in the general area of western North Dakota. Each of seven alternatives, with the exception of dry pressed brick, is examined for preliminary economic feasibility, based on information available locally, for utilization in the general area of Beulah and Bismarck, North Dakota. In each case where practical, a basis for calculations of 100,000 tons/year of fly ash or gasifier ash was used. Although many variations in process technology are possible for each of the alternatives, a single process to a single product was selected for each of the seven alternatives for economic analysis. A simple form of the product was selected where possible. For example, the mineral wool product selected was baled or chopped mineral wool rather than a finished product such as insulation batting.

The costs are established with the following general assumptions:

1. The fly ash and gasifier bottom ash are turned over to a separate organization for recovery or reuse at no cost. The cost offset to

the power or gas producer is the elimination of the cost for disposal. Some of the transportation costs prior to burial would probably not be recovered because of the need to move the waste to a transfer or backing site. Estimated disposal costs for the Antelope Valley station are presented in Table I. Initial burial costs of $5.00/ton and up are projected to drop over time to à little less than $4.00/ton in later years.

2. The ash utilization group pays for transportation of the ash to the use point, with the expectation that the plant location is in some reasonable proximity to both the utility producer and a point of ultimate use of the products. The location of a real production facility is a marketing decision and is not part of this study.

3. The organization utilizing the waste material is independent of the utility producer and does not receive low cost energy or free waste heat. Energy costs are based on usual industrial rates.

TABLE I

ANTELOPE VALLEY STATION SOLID WASTE TRANSPORTATION AND DISPOSAL COST FEASIBILITY STUDY ($ x 10³, TONS x 10³)* [2]

	1984	Total through 1992
Transportation		
Capital Invest. (7-1-83)		
Fixed Charges		
Interest 12%	1373.7	13050.2
Deprec. 3.10%	354.9	3371.6
Admin. 0.25%	28.6	271.7
Variable Costs		
Power	118.2	2179.2
Oper. Labor	25.6	371.7
Maint. Labor	8.5	122.7
Maint. Mtrls.	20.0	271.6
Total Solid Waste		
Transportation Cost	1929.5	19638.7
Cost ($)/Ton Trans.	8.97	4.11
Tons/yr of Solid Waste	215.1	4778.3
Disposal		
Variable Costs		
Site Prep.		
Liner Installation	49.5	1285.6
Reclamation	57.8	2460.6
Operation	69.7	2189.0
Environmental	10.0	244.8
Administrative	4.7	204.0
Total Solid Waste		
Disposal Cost	191.7	6384.0
Cost ($)/Ton Disposal	0.89	1.33
Cost ($)/ton T & D	9.86	5.44

*Based on 1981 cost estimate.

TABLE II

ANTELOPE VALLEY ASH ANALYSES (NORMALIZED)*

	Original Fly Ash	Scrubber Ash**	Gasifier Ash***
SiO_2	28.0	23.7	25.7
Al_2O_3	12.5	10.4	12.5
Fe_2O_3	7.3	5.1	16.0
CaO	24.7	28.9	23.7
MgO	7.8	6.4	5.9
Na_2O	9.4	9.1	12.1
K_2O	0.5	0.5	0.5
SO_3	9.8	15.9	2.7
TiO_2	n.a.	n.a.	n.a.
Total	100.0	100.0	99.1
Acid-Base Ratio	1.25	0.97	1.29
Sample #	CBUI 83-342	CBUI 83-523	GF 83-2247

*UNDERC Data
**Solid product of dry scrubber in which fly ash and lime are the active materials.
***METC gasifier operated on Beulah-Zap lignite.

4. Costs are based on average ash analyses (see Table II) for original and scrubbed fly ash and for ash obtained from a gasifier using Beulah-Zap lignite.

Details of the assumptions for each product are given in an appendix at the end of this paper.

OPTIMUM REPLACEMENT OF PORTLAND CEMENT WITH ASH

The substitution of fly ash for the cement portion of ready-mix concrete is technically practical and economically feasible. Fly ash has long been used for this purpose in other parts of the United States, and lignite fly ash has been found to be eminently satisfactory. The fly ash may be introduced at a number of points in the production of cement and its conversion to concrete. The application treated here is its use as a direct replacement for cement in ready-mix concrete.

With present practice, it is quite feasible to substitute 25% of the cement in a cubic yard of concrete with lignite fly ash, and the fly ash generated by the Antelope Valley station would be quite suitable for this purpose.

Although fly ash can be used as a direct ingredient in cement manufacture, the lack of availability of limestone in North Dakota makes a commercial cement plant seem unattractive. However, fly ash could be readily used in admixture with cement for ready-mix concrete and other concrete products such

TABLE III

FLY ASH REPLACEMENT IN READY-MIX CONCRETE

Raw Material Requirements and Cost/Cu. Yard Concrete

| | No Fly Ash Added | | 25% Cement Replacement | |
	lb/cu yd	$/cu yd	lb/cu yd	$/cu yd
Cement	465	$19.76	350	$14.88
Sand and Gravel	3350	12.56	3135	11.76
Fly Ash	0	--	160	0.00
Additives	--	6.30	--	6.30
Total		$38.62		$32.94

Total Cost/Cu. Yard Concrete

	No Fly Ash Added	25% Cement Replacement
Capital	$ 3.22	$ 3.22
Raw Materials	38.62	32.94
Electricity and Fuels	2.59	2.59
Labor	14.13	14.13
Pretax Profits	2.59	2.59
Transportation from power plant	6.43	6.43
Total	$67.58	$61.90

For 100,000 TPY Fly Ash Used

Cement used with fly ash	219,000 TPY
Yards of concrete produced	1,200,000 cu yd
Cost without fly ash	$84,475,000
Cost using fly ash (25% replacement)	$77,375,000
Saving	$7,100,000
Savings, $/cu yd of concrete	$5.68
Savings, $/ton of fly ash used	$71.00
Savings, % of cost without fly ash	8.4%

as sand-and-gravel blocks, lightweight aggregate blocks, and precast concrete pipe.

As an example of this use, costs have been estimated for the use of 100,000 TPY of fly ash in ready-mix concrete in the Bismarck area, largely updating costs from the 1980 Research Planning Associates, Inc., report [3]. These figures (see Table III) indicate that fly ash could effect a $7,100,000 savings at a 25% substitution rate, or about 8.4% of the cost of concrete without fly ash, amounting to $71 saved per ton of fly ash substituted.

MINERAL WOOL FROM ASH

Mineral wool is a fibrous material, made from rock and/or melted slag, and used in building insulation in the form of batts, board, granulated nodules or loose fibers. It is durable, fireproof, and insect- and vermin-proof. It is currently produced from waste slag obtained from steel mills. Bottom ash from cyclone or other "wet bottom" boilers has been proposed and tested as a substitute raw material for blast furnace slag, and gasifier ash would be very similar, physically and chemically, to bottom ash and consequently an attractive raw material for mineral wool.

The manufacturing process would consist of the following steps: 1) Balling of gasifier ash, 2) drying of ash pellets, 3) melting with coke in cupola, 4) spinning on disc into mineral fiber, 5) removal of shot on a shot screen, 6) granulation of spun product, and baling for shipment.

Gasifier ash for mineral wool manufacture would be collected in a granulated form from the gasifier discharge pool. The ash would be formed into balls approximately 2 inches in diameter on a balling drum, dried to give green strength, and fed alternatively with coke at a 3:1 ratio into a cupola in which the ash is melted at 2800°-3000°F. A molten stream of ash is fed from the cupola to a spinning device which spins the material into mineral fibers. The product contains larger particles, above 325 mesh screen size, which are removed by shot screens. The shot can be recycled to the balling step. Fibers 5 to 7 microns in diameter are produced, and these are granulated for packaging and bulk shipment. For best operations the "acid-base ratio," i.e., silica plus alumina to calcium oxide plus magnesia ratio, in the ash should be from 0.8 to 1.2 to facilitate melting and fiberization. Lignite scrubber ash from the Antelope Valley station was found to have an acid-base ratio of 0.97, well within the acceptable range.

An economic summary for the manufacture of mineral wool from 100,00 TPY of ash in western North Dakota is presented in Table IV. The plant investment cost is estimated at $2,109,000 and the annual production cost at $12,488,000 or 9.61¢/lb of product. Current selling price of 13.5¢/lb indicates that a comfortable profit margin is possible.

BLENDED CEMENT UTILIZING ASH

The most promising potential bulk application of fly ash in direct cement manufacture is the LTM (Lient Trief Mixed) process [5], developed in Belgium, which utilizes about 70% of fly ash to produce a material comparable to portland cement.

The first step in manufacture is the production of LTP (Lient Trief Pure), a binder containing 50% fly ash and 50% lime, by melting 40% fly ash and 60% limestone at 1430°C, followed by quenching in water and crushing to below 10 microns. The LTP is blended at a ratio of 40% LTP to 60% fly ash to produce LTM 300, which is the equivalent of P300 portland cement.

Properties claimed for the material are shorter curing time, more complete hydration of the cement, less heat generation in the curing process, and superior resistance to saline water as compared to concrete from portland cement. The material can be stored wet, since the cementing reaction does not begin until a caustic soda catalyst is added, and this addition may take place at a ready-mix plant.

The process mode selected for cost analysis here is the manufacture of LTM 300 at a western North Dakota location close to the ash source. Since caustic soda must be added to the LTM as a catalyst for the setting reaction, and the end use here would be mainly in ready-mix concrete, it was decided to add the caustic dry to the LTM mix, and store dry as with conventional cement, to cause the least change in practice at the ready-mix plant.

TABLE IV

MINERAL WOOL MANUFACTURE

Basis: 100,000 TPY Fly Ash or Gasifier Ash

Raw Materials	TPY	Cost/Ton Value	Annual Cost/Value
Fly Ash	100,000	0.00	--
Coke	33,300	$105.00	$ 3,496,000
Product (13.5¢/lb)	65,000	270.00	17,555,000

Total Product Cost

Item	% of TPC**	$/yr	¢/lb Prod.
Raw Materials	28	3,496,000	2.69
Operating Labor	13	1,623,000	1.25
Supervision	2	250,000	0.19
Utilities	10	1,249,000	0.96
Maintenance and Repairs	3	375,000	0.29
Operating Supplies	1	125,000	0.10
Lab., Royalties	3	375,000	0.29
Plant Overhead*	10	1,249,000	0.96
Administration	5	624,000	0.48
Distribution and Selling	18	2,248,000	1.73
Research and Development	5	624,000	0.48
Financing	2	250,000	0.19
	100	12,488,000	9.61

Plant Investment $ 2,109,000

Current cement selling price bagged mineral wool $0.135 per lb.

---

*Includes general plant upkeep and overhead, packaging, storage
 facilities, safety, and protection.

**Estimated by procedures in Peters and Timmerhaus [4].

Table V presents a preliminary economic analysis of a TRIEF plant located
in western North Dakota and using 100,000 TPY of ash to produce 136,000 TPY of
LTM containing 8% dry caustic. Conventional plant costs for comparison are
given in the Resources Planning Associates report of June, 1980 [3]. For this
plant, as for any potential cement plant in North Dakota, the necessity of
shipping in limestone (and in this case caustic) causes a very high raw
material cost. Offsetting this is the much lower capital cost of a TRIEF
plant, together with its lower energy requirements because less of the total
material is heated.

HIGH FLEXURAL STRENGTH CONCRETE

Lignite fly ash can be used as a principal raw material in the manufacture
of ceramic products, and has been proposed in the production of power poles,
fence posts, and power line insulators. A study by EPRI [6] published in 1980

TABLE V

TRIEF CEMENT MANUFACTURE

Basis: 100,000 TPY Fly Ash or Gasifier Ash

Raw Materials	TPY	Cost/Ton	Cost/Year	Cost/Ton Product
Fly Ash	100,000	0	0	0
Limestone	37,500	$ 36.20	$1,357,000	$ 9.98
Caustic Soda	11,000	122.00	1,342,000	9.87
Total	148,500		$2,699,000	$19.85

Product   136,000

Total Product Cost

	TRIEF Cement			Competitive Conventional Cement
	% of TPC	$/yr	$/Ton	$1 Ton
Raw materials	37.6	2,700,000	19.85	7.18
Capital Cost	7.3	477,000	3.53	18.80
Electricity	8.8	575,000	4.28	6.41
Fossil Fuel	18.4	1,202,000	8.92	14.88
Labor	27.9	1,824,000	13.47	13.47
Total Product Cost	100.0	6,778,000	50.05	60.74
Selling Price			53.31	64.00
Transport. to ND				21.00
Delivered Price			53.31	85.00

	TRIEF	Conventional
Plant Cost	$5,657,000	$30,200,000
Capacity, TPY	136,000	136,000
Product Cost/Ton	$ 50.05	$ 60.74
DELIVERED PRICE	$ 53.31	$ 85.00

gave considerable information on a conceptual operation for power pole manufacture based in California. In this report, the EPRI data have been modified to assess a North Dakota operation in which power poles are made from 59% fly ash, 9% clay, and 32% waste glass. The study by Resources Planning Associates, Inc., for the Mercer County, North Dakota Energy Development Board [3] pointed out that the availability of waste glass in the North Dakota area is not sufficient to support a commercial operation. Use of waste glass brokers from other parts of the Midwest is an alternative, but high collection and shipping costs are an economic disadvantage. Crushed glass, clay, and fly ash are blended, extruded wet, formed, dried, and fired in a kiln to produce the final product.

A cost estimate based on EPRI estimates updated to 1984, scaled to 100,000 TPY of fly ash, and with raw material costs estimated for western North Dakota, are shown in Table VI. The indicated cost per pole is less than

TABLE VI

PRODUCTION COST FOR POWER POLES FROM ASH

Basis: 100,000 TPY Gasifier Ash

Raw Materials	% in mix	TPY	$/Ton	$/Ton Product	$/Pole
Clay	9	15,300	$ 15.00	$ 1.35	$ 1.96
Waste Glass	32	54,200	105.00	33.57	48.68
Gasifier Ash	59	100,000	0.00	0.00	0.00
Total	100	169,500		$34.92	$50.64

Production Cost of 40-ft Power Pole

	Cost/Ton Products	Cost/Pole
Capital	$ 5.12	$7.43
Raw Materials	34.92	50.63
Electricity and Fuels	12.73	18.46
Labor	11.32	16.41
Maintenance and Administration	6.92	10.04
Overhead and Plant	2.66	3.86
Total	$73.67	$106.83
Corrected for 90% Yield	81.86	118.70

40-ft Power Poles (2900 lb/pole):

Annual Production Cost	$12,487,000		
Annual Production, Tons	169,500		
Annual Production, Pole	116,900	90% Yield	105,200
Cost/Ton Product	$ 81.86		
Cost/Pole at 90% yield	$131.89		
Wood Poles (NSP):	60' Class 2	$318	
	45' Class 2	$252	
Savings/Pole	$134		
Savings/Year	$14,100,000		

half of a comparable wood pole, indicating that the process should be considered further.

UTILIZING ASH TO CONSTRUCT HAUL ROADS

Ash usage in cement and road building in North Dakota is not new, but full usage has not been reached in either application. There are many methods, proposed and in use, of incorporating fly ash or bottom ash materials in road construction. The application evaluated in this study is to use ash to upgrade existing unpaved county roads by providing paving and increased stabilization. Road construction using a 4 inch hot mix mat over a gravel base is estimated at $122,0000/mile. Reworking the old mat and including 50% new hot mix is $81,400/mile.

TABLE VII

SECONDARY ROAD CONSTRUCTION USING ASH

LIME - FLY ASH STABILIZATION OF EXISTING ROAD BED:

Cost estimates are based on $/mile for 1 inch deep surface mat, 28 ft wide, on 32 ft wide, 8 inch deep base.

Construction volume of road bed 112 696 ft^3/mile; finished specific gravity: 2.4

Raw Materials	Amount		Total Cost
Lime 2%	168.85 ton	$96/ton	$16,210
Fly Ash 12%	1013 ton	--	--
Primer for surface			
for 28 ft wide	3285 gal	0.95/gal	3,121
Seal Oil	6568 gal	0.80/gal	5,254
Capping Bituminous Surface	1010 ton	15.20/ton	15,352
Cover Coat Blotter Material	80 ton	6/ton	480
Total Raw Material			$40,417

Construction	
Bed Preparation	$ 5,000/mile
Raw Materials	40,417/mile
Equipment Usage and Maintenance*	8,000/mile
Labor	15,000/mile
	$ 68,417/mile

OTHER OPTIONS:

New road hot mix on gravel base	
Gravel	$ 5,000 mile
Mat	100,000 mile
Labor	17,000 mile
	$122,000/mile

50-50 old mat mixed with new hot mix	
Prep	$ 5,000/mile
Hot mix	61,400/mile
Labor	15,000/mile
	$ 81,400/mile

---

*Often rented or leased

The method evaluated in this report involves stabilizing an existing road bed by breaking up the existing mat, usually 4 inches thick with 4 inches of clay and gravel underneath, and incorporating 2% lime and 12% fly ash. The width of road treated is 32 feet. A 28 foot wide mat of one inch thick hot mix asphalt is used to cover the lower stabilized portion.

Table VII presents cost estimates and indicates a cost per mile of $68,700 versus $122,000 for new hot mix on a gravel base.

Civil Engineering bids and estimates usually include labor for installation with each item on the materials list. Labor has been totaled and listed separately in this study. Equipment used in this method of road construction requires no differences in maintenance.

CERAMIC GLAZED WALL TILE USING ASH

Wall tile is generally 4.25 in x 4.25 in for the glazed surface. The back is patterned for ease in setting and spacer feet are formed on the edges to establish a uniform grout space. Production and sales are usually costed per square foot as is done in this study. Wall tile is generally formed in a press and is dried and the glaze is applied while the tile is green (unfired). Glaze application is usually done by spraying. The tile is fired only once. Unglazed tile requires a fired clay body with little or no water absorption. Wall tile has acceptable water absorption of up to 18%. Glaze costs are highly variable and difficult to establish for a clay body with some color. Estimates for a tile plant using a three hour fire to cone 01 (about 2000°F) have been used. The tile in the study is expected to have a retail list price of $2.40 to $3.00 per square foot. The price of the tile as it leaves the facility is approximately $1.06 per square foot. Transportation costs add a great deal to the final price.

Wall tile is the only product in the study produced in a form usable by a consumer. Marketing costs are difficult to estimate in dollar terms for such a product. Of concern with the tile, but not a problem in other ash uses, is quality control and loss of production due to breakage, warpage, and nonuniformity in body, glaze color, and texture. This concern results in differences between design capacity of the plant and actual production of an estimated 20%. For ease in calculations, weights have been reported on a dry basis.

Table VIII shows estimated raw material costs and product costs for a plant producing a net 571,000 square feet of tile per year, using 950 TPY of fly ash. Raw material costs are seen to be substantially less with fly ash utilization, resulting in a savings of about 5.5¢/tile or 5.2% of the sales price.

SULFURCRETE® UTILIZING ASH

Sulfurcrete® is a concrete product utilizing plasticized sulfur as a 100% replacement for portland cement. It was developed by Sulfur Innovations, Ltd., of Calgary, Alberta, Canada, who hold the patent [7].

Sulfurcrete® is manufactured by mixing a preheated, preblended, dry aggregate mixture with melted sulfur. A proprietary SRX plasticizer is then blended into the mixture at about 270°F. After ten minutes of reaction, the material can be poured into molds and immediately tamped or vibrated for air pocket removal and consolidation. The mixture solidifies quickly and can then be removed from the mold.

The cast Sulfurcrete® reaches compressive strengths of 6,000 psi to 10,000 psi in two or three days. It is unaffected by sulfates, salts, and most acids. It is nearly impervious to water, and, therefore, resistant to freeze-thaw damage. It can be repeatedly loaded to 95% of its ultimate strength with no evidence of fatigue, where cured portland cement concrete will exhibit fatigue after repetitive loading to 60% of its ultimate strength.

Sulfurcrete® has been used in Alberta for many precast products including parking curbs, highway medium barriers, traffic islands, support beams, and elevator counter weights. It can also be used in field-poured applications.

TABLE VIII

WALL TILE MANUFACTURE

Basis: 950 TPY fly ash in a plant designed for 714,000 ft^2/yr production

Breakage is estimated to be 20%. 571,000 sq ft of tiles produced.

Raw Materials	% Used	TPY	Cost Cost/Ton	Using Ash	Conventional
Clay	20	190	$15	$ 2,856	$ 2,856
Feldspar	10	95.2	65	--	6,188
Flint	70	666	18	--	11,995
Fly Ash	80	951	--	--	--
Subtotal for Clay Base				2,856	21,039
Glaze				21,420	21,420
Total				$ 24,276	$ 42,459
Total Capital Cost				$600,000	$600,000

Total Product Cost/Square Foot
Total Design Production Before Breakage

	Fly Ash Used $ per sq ft	Conventional $ per sq ft
Raw Materials		
Clay base	0.0040	0.0295
Glaze	0.03	0.03
Capital Equipment	0.084	0.084
Electricity and Fuel	0.14	0.14
Labor	0.17	0.17
Transportation	0.05*	0.07**
Water, Mold Release and Other Supplies	0.02	0.02
	0.498	0.5435
Breakage	0.0996	0.1087
Cost of Production of Good Tile	$ .5976	$ .6522
Sale Price of Tile	$ 1.06	$ 1.06

*Local sources of all product.
**No local source of feldspar.

Studies have been made at the UND Engineering Experiment Station on the use of lignite gasifier bottom ash as aggregate in Sulfurcrete®; a very successful mixture comprised of 72.5% crushed gasifier ash, 25% sulfur, and 2.5% SRX plasticizer.

Estimated costs for Sulfurcrete® at present conditions in western North Dakota are shown in Table IX. Since the overwhelming factor in costs is that

TABLE IX

ECONOMICS OF SULFURCRETE®

Basis: 100,000 TPY Gasifier Bottom Ash

Raw Material Cost

Raw Material	% in Product	TPY	$/Ton	$/cu yd
Gasifier Ash	72.5%	100,000	0.00	0.00
Sulfur	25	34,480	$ 133.00	$ 55.65
Plasticizer	2.5	3,450	1120.00	46.90
Total	100.0	137,930	$1253.00	$102.55

For 100,000 TPY Gasifier Ash Used

Yards of Sulfurcrete®, produced	82,395
Cost of annual production Sulfurcrete®	$ 10,835,800
Cost of equal volume portland cement concrete	$ 5,568,300
Excess cost of Sulfurcrete®	$ 5,267,500
Cost/cu yd of Sulfurcrete®	$ 131.51
Excess cost/cu yd of Sulfurcrete®	$ 63.93

of raw materials, no attempt was made to differentiate between Portland cement concrete and Sulfurcrete® in other than raw material costs.

SUMMARY AND CONCLUSIONS

A summary of the preliminary economics of the seven alternative applications is shown in Table X. The expected cost savings range to a high of 52.8% for replacing wood power poles with high flexural strength concrete poles made from 59% gasifier ash. Not taken into account was the possibility that the concrete poles might have a longer life expectancy than the wood poles; such a result would improve the advantage even further. Sulfurcrete® from ash was found to be 94.6% more expensive than regular concrete; its superior chemical resistance and other properties might, however, justify its use in certain applications. The case of mineral wool from ash did not provide a good comparison base; the competitive raw material would be blast furnace slag which would have to be shipped a long distance for use in the North Dakota area.

This report is based on very approximate estimating procedures. Assumptions made in estimation are presented in the next section. A second phase, with a more precise economic assessment, would require much more precise knowledge than is currently available of process details, equipment costs, and information on the exact analysis and physical properties of the ash produced by the power plant or gasifiers of interest. At the time of writing, no ash from the gasifier was yet available. Fly ash from baghouses or precipitators has been used in most experimental work to date; dry scrubber ash should also be evaluated. It should be preceded by a market survey to assess realistically the plant sizes proposed.

TABLE X

COMPARISON OF ALTERNATE ASH APPLICATIONS

Application	Basis TPY Ash	Product	Units	Prod/Year	Cost/Year With Ash	Cost/Unit With Ash	Cost/Unit Without Ash	Percent Savings
1. Ash mix concrete (cement replacement)	100,000	Concrete	cubic yard	1,200,000	$77,375,000	$ 61.90	$ 67.58	8.40%
2. Mineral Wool	100,000	Granulated wool in bales	ton	65,000	12,488,000	9.61	-*	-*
3. Cement Manufacture (TRIEF)	100,000	Replacement Cement	ton	136,000	6,778,000	53.31	85.00	37.3%
4. Sulfur-crete using ash	100,000	Poured Sulfurcrete	cubic yard	82,400	10,835,800	131.51	67.58	(94.6%)**
5. High Flexural Strength Concrete	100,000	Power poles (2,900 lb ea.)	no. of poles	105,200	12,487,000	118.70	252.00***	52.8%
6. Ash in Haul Roads	1,013	Road surface (32' wide, 8" deep	Mile	-	-	68,417	122,000	44.4%
7. Wall Tile	950	Glazed wall tile tile	square feet	714,000	340,200	0.597	0.652	5.2%

* Alternative is blast furnace slag, not a practical alternative in North Dakota
** Excess cost would be justified by improved properties
*** Wood poles

APPENDIX: Assumptions made for economic estimations.

Replacement in Ready-mix

1. Fly ash can be used to replace 25% of cement in ready-mix.
2. Costs (re: Bismarck contractors): Cement $85.50/ton, sand $5.50/ton, gravel $10.50/ton, fly ash $25.50/ton, admixtures (air entrainment agent and water reducer) $6.30/cu yd, concrete (ready-mix) $63.10/cu yd.
3. Fly ash is available at no cost to the user f.o.b. the GPGA site (or Bismarck); assumed for all applications.
4. Ready-mix composition and requirements taken from report "Economic Feasibility of Using Fly Ash in the Ceramic and Concrete/Cement Industries", Contract No. DE-FG03-80CS20028 [2].
5. All cost differences in ready-mix concrete, with and without fly ash, are differences in raw material costs; other costs are assumed the same for both.
6. Sand and gravel are added in the proportion, 40% sand and 60% gravel.
7. Raw material costs taken from (2) above. Other costs obtained by extrapolative breakdown from the report in (4) above to the total current cost given in (2) above.

Mineral Wool

1. Coke cost--$105.00/ton, from O. Manz.
2. Plant cost--Marshall and Stevens equipment index used to obtain 1984 costs.
3. 0.6 power used for scaling up or down in plant size from furnished information.
4. Mineral wool plant cost--$1,000,000 for 4 ton/hour plant (Manz); scaled to 100,000 TPY.
5. 35% of mineral wool is discarded as shot.
6. Procedures of Peters and Timmerhaus [4] used for product costs, based on plant and raw material costs.

TRIEF Cement

1. Material requirements as given by Trief [4].
2. Caustic soda bulk pellets 61¢/lb (March 1984, Chemical Marketing Reporter).
3. Amount of caustic (8% of dry LTM) from experiments of O. Manz.
4. Caustic can be blended dry, material measured in sacks and water added at ready-mix plant or point of use.
5. Costs modified from conventional cement plant costs [3].
6. Limestone costs f.o.b. Bismarck, ND, obtained from O. Manz.
7. Electricity and fuel costs assumed 1/3 lower than conventional because less material is heated.

Sulfurcrete®

1. Costs other than raw materials assumed same as conventional portland cement concrete.
2. Sulfur cost $133/ton current at gulf ports (Jan. 1984 E. & M.J.). $55/ton is probably nearer to what GPGA expects to obtain for its Stretford sulfur, but the higher cost was used in the calculations.
3. Plasticizer assumed to be $1120/ton. Latest price quotation was $ Can. .75/lb, which is at current exchange a little less than $1200/ton in U.S. funds.

High Flexural Strength Concrete

1. Proportion of gasifier ash 59%, clay 9%, waste glass, 32%.
2. Costs modified from those presented in EPRI report.
3. Wood pole prices obtained from NSP.
4. Breakage of 10% of poles is assumed (from EPRI estimate).
5. Standard pole same as EPRI report.

Road Construction Using Ash

1. Cost estimated on the basis of discussions with Oscar Manz and information on road costs in Oliver Co., ND.

Glazed Wall Tile Using Ash

1. Clay body formulation and firing conditioning based on information in sections on ceramics from "Encyclopedia of Chemical Technology" Interscience Publishers, New York, NY, 1948.
2. Firing and energy calculations estimated on the basis of material in "Handbook on Material and Energy Balance Calculations in Matallurgical Processes." H.A. Fine and G.H. Geiger, The Metallurgical Soc. of AIME Warrendale, PA, 1979.
3. Three unfired tiles per pound of raw tile mixture.
4. The 20% breakage is an estimate based on tile factory processing figures and profitability.

ACKNOWLEDGMENTS

Supported by GRI Contract No. 5082-235-0771 through the North Dakota Mining and Mineral Resources Research Institute.

REFERENCES

1. Manz, Oscar E. "Utilization of By-Products from Western Coal-Combustion in the Manufacture of Mineral Wool and Other Ceramic Products," Cement and Concrete Research, 14, 513-519 (1984).

2. Communication: Ericksen, R.L. Basin Electric Power Cooperative to G. Groenewold, NDMMRD, December 20, 1983.

3. Economic Feasibility of Using Fly Ash in the Ceramic and Concrete/Cement Industries. Report prepared for Energy Development Board of Mercer County, North Dakota, by Resources Planning Associates, Inc., Washington, D.C. (From U.S. Department of Commerce, Economic Development Agency, Project No. 05-06-0-1878, and U.S. Department of Energy, Buildings and Community Systems Division, Contract No. DE-FG03-80DS20028). 112 pp.

4. Peters, Max S. and Timmerhaus, Klaus D. Plant Design and Economics for Chemical Engineers. 2nd Edition, McGraw-Hill, New York, 1968, pp. 140-41.

5. Trief, L. The New TRIEF Cement Process. International Ash Symposium. Atlanta, GA, 1979.

6. Development of Power Poles from Fly Ash. Phase II. Project 851-1 Final report, ECP, Inc., El Segundo, California for EPRI, September 1979.

7. Vroom, Alan H. U.S. Patent 4, 058, 500, November 15, 1983.

LEGAL NOTICE

PART IV

Environmental Considerations

DISPOSAL OF WESTERN FLY ASH IN THE NORTHERN GREAT PLAINS

GERALD H. GROENEWOLD*, DAVID J. HASSETT**, ROBERT D. KOOB***, and
OSCAR E. MANZ****
  * North Dakota Mining and Mineral Resources Research Institute, Box 8103,
    University Station, University of North Dakota, Grand Forks, ND 58202
 ** Fuels Analysis Lab, Engineering Experiment Station, Box 8103, Univer-
    sity Station, University of North Dakota, Grand Forks, ND 58202
*** College of Science and Mathematics, North Dakota State University,
    Fargo, ND 58102
**** Coal By-Products Utilization Institute, Box 8115, University Station,
    University of North Dakota, Grand Forks, ND 58202

(Received 29 November, 1984; accepted 25 February, 1985; Refereed)

ABSTRACT

Leachates from western fly ashes are typically alkaline. Our studies
indicate a strong correlation between alkalinity of western fly ash leachate
and trace element concentrations. Elements of particular concern include
As, Se, and Mo. A base neutralization mechanism is operative in all of the
overburden types found at mine disposal sites in western North Dakota.
Regional geological similarity suggests that this mechanism is operative
throughout the Northern Great Plains. Although the mechanisms of neutralization are speculative, laboratory experiments indicate significant neutralization at all levels of base above background levels. Long-term
monitoring of fly ash disposal-sites indicates that alkaline neutralization
of fly ash leachate is occurring. Further, field data indicate that toxic
trace elements (particularly As and Se) in disposal site leachates decrease
in concentration as the pH of the leachate is neutralized. Thus, the
intrinsic conditions at Northern Great Plains fly ash disposal sites appear
to promote significant attenuation of critical toxic elements found
in fly ash leachates. Regardless of the pH, leachates in those settings
have high concentrations of sodium and sulfate. Western fly ashes are
commonly cementitious. Our studies indicate that fly ashes commonly develop
significant strength after several months of burial, particularly if emplaced in an unsaturated disposal setting. Once cementitious reactions have
occurred, the fly ashes show little potential for leaching. Thus, a combination of intrinsic disposal-site conditions and the cementitious behavior
of the fly ashes suggests that surface-mine disposal of western fly ashes in
the Northern Great Plains, assuming proper disposal-site selection, may not
cause long-term environmental problems associated with toxic trace elements.

INTRODUCTION

Increased use of coal for production of electricity generates large
volumes of solid wastes, mainly in the form of fly ash and flue gas desulfurization (FGD) waste as residues from emission control systems. Placement
of these wastes in surface mine areas and backfilling with overburden is a
common method of disposal at power plants located next to surface coal
mines. Although this practice has existed for many years, little is known
about the effects of fly ash and FGD waste upon the quality of groundwater
in and near the disposal areas.

Fly ash and FGD waste constitute the two major solid by-products of
lignite-burning power plants in North Dakota. Approximately 1,750,000 tons
of coal conversion solid wastes are generated annually in North Dakota.
These quantities will increase significantly as additional thermoelectric
and lignite gasification facilities are constructed. Of concern is the
generation, and potential impact on groundwater, of highly mineralized

leachate having very high concentrations of sodium and sulfate. Also of concern is the possibility of generating leachate with elevated concentrations of certain potentially toxic trace elements, particularly arsenic, selenium, and molybdenum.

This paper summarizes results obtained to date from an ongoing investigation, initiated in 1978, of the potential effects of fly ash and FGD waste on groundwater quality at the Center Mine, operated by Baukol-Noonan, Inc., near the town of Center in western North Dakota (Figure 1). This paper focuses specifically on the research related to disposal of fly ash. This research involves detailed field studies, laboratory studies, and computational geochemical studies. Previous laboratory studies involved column and batch leaching experiments.

The waste is generated by the Square Butte Electric Cooperative (SBEC) Milton R. Young Station, a mine mouth power plant that fires low sulfur-lignite coal from the Center Mine. The station consists of two cyclone-fired generating units; Unit #1 is a 240 MW boiler, and Unit #2 is a 450 MW boiler. Both units are equipped with electrostatic precipitators (ESP), and Unit #2 is equipped with a wet scrubber FGD system that uses the alkalinity in the fly ash in lieu of lime/limestone as the sulfur capture agent.

The field study is evaluating two hydrogeologically distinctive disposal settings in the Center Mine: the pit bottom and the vee-notch between spoil ridges. In the restored landscape, pit-bottom disposal sites are typically below the water table (a saturated setting) and have relatively high permeabilities. The vee-notch sites are typically above the reestablished water table (an unsaturated setting) and are commonly enclosed by materials of relatively low permeabilities. Subsurface water monitoring instrumentation has allowed us to evaluate baseline groundwater conditions, the effects of fly ash on groundwater quality in the area of the Center Mine, and surface mine disposal-site design criteria.

Previous laboratory studies and field monitoring at the Center, North Dakota disposal sites indicated that North Dakota fly ash typically generates a highly mineralized and very alkaline leachate (pH = 10 - 12.5). Under these conditions, arsenic and selenium are apparently highly mobile and often reach potentially toxic concentrations in groundwater that is in contact with the fly ash [1]. Elevated concentrations of molybdenum have also been observed in the fly ash leachates. These observations have generated particular concern regarding the design of fly ash disposal sites in North Dakota.

Previous research on the impacts of surface mining on groundwater at several North Dakota surface mine sites [2] has suggested that an acid/base buffering mechanism, which inhibits acid production and behaves like an acid toward high pH solutions, is operative at all the western North Dakota sites examined to date. Such a mechanism, if typical of North Dakota overburden, could be extremely significant to trace metal attenuation of fly ash leachate, particularly if arsenic and selenium mobility are related to alkalinity. In turn, if such a mechanism is operative at all the western North Dakota disposal sites, the need for expensive liners could be significantly decreased.

Two types of laboratory studies have been conducted. The first has focused on a detailed evaluation of the mechanisms and ability of all the types of sediments present at North Dakota surface mines to neutralize alkaline fly ash leachate. The second has focused on the ability of these sediments to attenuate selected trace elements. Data from these laboratory studies, in conjunction with long-term monitoring of fly ash disposal sites, is allowing us to evaluate the intrinsic trace element attenuation mechanisms operative in western North Dakota disposal sites. In turn, these data

FIGURE 1. LOCATION OF CENTER, NORTH DAKOTA STUDY AREA

will provide an objective basis upon which to design fly ash disposal sites in North Dakota surface mines and, particularly, allow us to evaluate the need for costly liners.

METHODOLOGY

Subsurface Water Instrumentation, Monitoring, and Sampling

Approximately 250 piezometers have been installed in and around the Center Mine. Approximately three-fourths of these are screened either in or below the spoils. The remaining piezometers are screened in stratigraphically equivalent positions in undisturbed areas next to the mine. In addition, 12 nests of pressure-vacuum lysimeters have been installed; 10 in the vee-notch areas and 2 in undisturbed sediments next to the mine. The pressure-vacuum lysimeters were installed to monitor subsurface water quality in the unsaturated zone.

All piezometer test holes were drilled using a rotary rig. Virtually all the holes were drilled using only air for circulation, thereby eliminating potential chemical contamination from the injection of drilling fluids. Subsurface water sampling procedures followed EPA recommendations [3]. All water samples were obtained by bailing. Because of relatively low permeability of the spoils and undisturbed sediments at the study site, each piezometer was either bailed dry or a maximum of two water column volumes were removed before sampling. A more detailed discussion of well installation and sampling procedure can be found in Groenewold et al. [1,4].

Water levels in all the piezometers were monitored on a monthly basis. These data were used to interpret groundwater flow in the study areas. In addition, single-well-response tests of selected piezometers were conducted to determine the hydraulic conductivities of the spoils and various undisturbed units.

Alkaline Neutralization Experiments

Overburden materials from four North Dakota mine sites have been used in the alkaline neutralization studies. These sites are the Center, Indian Head, Falkirk, and Glenharold Mines in west-central North Dakota. Detailed discussions of these mine sites can be found in Groenewold, et al. [2]. Overburden samples were obtained with a truck-mounted hollow-stem auger using 7.5 x 75 cm Shelby tubes. Immediately after withdrawal of the Shelby tube, the ends were sealed with heavy tape to avoid contact with air. The Shelby tubes were then sent to the laboratory for sample extrusion. If the sample could not be extruded, a portion of the tube was sawed off to obtain a sample for experimentation. Unused portions of the samples were stored in the tubes with the ends resealed. Table 1 provides the sample numbers used for the alkaline neutralization experiment, the field designation of the sample, the approximate depth (in feet) of the sample, and a qualitative description of the type of overburden material in the sample.

TABLE 1

SAMPLE DESCRIPTION

Sample Number	Field Designation	Depth of Sample	Description
1	IH-X2	17'-20'	Reduced clay
2	Consol 31	2'-4', 14.5'-16'	Oxidized clay
3	Ctr 361	40'-45'	Reduced sand
4	Ctr 380	9'-12'	Oxidized sand
5	Fa 514	22'-27'	Reduced silt
6	Consol 31	4.5'-8'	Oxidized silt
7	Fa 576	64'-65', 69-70'	Reduced till
8	Fa 576	22.5'-25'	Oxidized till

Key: IH = Indian Head Mine; Consol = Glenharold Mine;
CTR = Center Mine; Fa = Falkirk Mine

In the laboratory, samples were removed from their plastic bags, ground to pulverize any consolidated material and oven dried overnight. Each sample was then subjected to a variety of treatments to measure neutralization capacity. Samples were neutralized with 0.17 M Ca(OH)$_2$ and 0.035 M, 0.1 M, 0.5 M and 1.0 M NaOH solutions. In each treatment, a two-gram aliquot of each material type was added to 25 ml of the previously standardized base contained in 125 ml Erlenmeyer flasks. The mixtures were then boiled for ten minutes, cooled for an additional ten minutes and then filtered with a glass fiber Micropore filter using a suction flask. The resulting clear solution was then titrated using a standard HCl solution to pH 7 using a Sargent-Welch digital pH meter as the indicator. In an attempt to avoid systematic error, two runs were made on each sample at each concentration of base with all samples done at one concentration before proceeding to the next. Then samples were run by subjecting them to all concentrations of base simultaneously; i.e., subjecting them to the same heating-cooling-filtering-titration cycle.

Comparable procedures were used on selected samples prepared for determination of silicon and aluminum except that Teflon beakers were used for solutions and the extract was stored in plastic bottles. Si and Al were determined using a Perkin-Elmer Model 603 Atomic Absorption spectrometer. This instrument carries graphite furnace and background correction options.

Total organic content for each material type was determined by placing 10 g of soil in a polyethylene beaker, decomposing carbonates with dilute HCl, removing silicates by treatment with HF, and then igniting the resulting dried material in a muffle furnace at 750°C. The resulting weight loss was attributed to organic matter.

Trace Element Attenuation Studies

Attenuation experiments have been carried out using selected overburden sediments typical of North Dakota fly ash disposal sites. The major objective has been to determine the cation and anion attenuation capacity of typical overburden materials from three surface mines in western North Dakota. Information has been generated on the attenuation of As, Se, Cd, and Mo. The two sediment types chosen were sandy silt and clay. These materials were evaluated in both oxidized and reduced states. Attenuation experiments were performed in which these materials were treated with solutions of selected cations and anions at known concentrations. The experiments were run at two ionic strengths. Five grams of sediment were shaken with 100 ml of solution in a wrist-action shaker. All experiments were run at room temperature. Typical ranges of parameters and conditions were:

$$
\begin{array}{lll}
\text{As} & 100 \text{ ppb} - & 4000 \text{ ppb} \\
\text{Se} & 100 \text{ ppb} - & 4000 \text{ ppb} \\
\text{Mo} & 200 \text{ ppb} - & 6000 \text{ ppb} \\
\text{Cd} & 9.2 \text{ ppb} - & 200 \text{ ppb}
\end{array}
$$

* Na_2SO_4 1479 mg/l   Low ionic strength

* Na_2SO_4 4473 mg/l   High ionic strength

* Ionic strength adjuster

These ranges and conditions were similar to those used by other workers [5,6,7]. NaOH was added in varying amounts, but because of the high neutralization capacity of these soils, pH change was minimal. This addition was described in terms of percent of theoretical neutralization.

RESULTS AND DISCUSSION

Groundwater Composition in Unmined Areas and in Spoils Without Waste

Chemical analyses of groundwater from undisturbed units and spoils not influenced by wastes at the Center Mine are summarized in Table 2. Although considerable variability in water chemistry is apparent between the various undisturbed units, all contain water that is generically very similar. The major ions in solution in all the undisturbed units are sodium, calcium, bicarbonate, and sulfate. The Kinneman Creek and Hagel lignite beds contain Na, Ca-HCO$_3$, SO$_4$ to SO$_4$, HCO$_3$-type water. Total-dissolved-solids concentrations in the Kinneman Creek and Hagel beds range from 843 to 1631 mg/l and 649 to 3874 mg/l, respectively (Table 2). The nonlignitic sediments underlying the Hagel lignite bed or spoils are characterized by Na to Na, Ca-HCO$_3$, SO$_4$ to SO$_4$, HCO$_3$-type water. Total-dissolved-solids concentrations in these units range from 416 to 5051 mg/l.

Sulfate concentrations in the undisturbed units commonly exceed the recommended limit for drinking water of 250 mg/l by a factor of 2 to 4. The concentrations of cadmium are generally well below the maximum permissible concentration of 10 μg/l. Selenium and arsenic are below the maximum permissible levels of 10 μg/l and 50 μg/l, respectively, in all samples. Recent data, not included in Table 2, indicate that background molybdenum

TABLE 2

CHEMICAL ANALYSES OF GROUNDWATER FROM SPOILS AND UNDISTURBED UNITS, CENTER MINE, CENTER, NORTH DAKOTA

	Field Temp °C	Field pH	Field Cond μmhos/cm	DO	TDS	Total Hardness	Ca	Mg	Na	K	HCO3 mg/L	CO3	SO4	Cl	NO3	F
SPOILS																
x̄	9.7	6.7	3880.0	0.8	3375.6	---	348.0	232.4	361.0	15.6	1165.4	---	1454.9	8.1	0.6	0.1
s	4.2	0.3	1243.2	0.4	910.7	---	97.5	90.1	255.1	8.9	413.1	---	536.7	4.6	0.4	0.1
n	21	21	21	15	16	---	16	16	16	16	18	---	18	18	13	15
high	16.5	7.7	8000	3.5	5338.0	---	519.0	431.2	1118.0	34.0	2113.0	---	2520.0	21.0	1.8	0.3
low	2.0	6.3	1830	0.2	1599.0	---	190.0	103.5	89.5	7.5	363.6	---	565.8	2.3	0.1	0.0
KINNEMAN CREEK BED																
x̄	10.4	6.7	1655.6	0.4	1229.3	404.3	84.3	51.9	249.7	18.8	651.1	1.1	372.2	8.1	1.8	0.1
s	2.8	4.6	437.1	0.2	333.3	227.0	47.0	20.3	122.7	11.3	208.7	0.4	137.5	8.0	2.6	0.0
n	9	9	9	5	9	4	9	9	9	9	9	3	9	9	13	6
high	16.0	7.5	2200	0.7	1631.0	731.0	165.0	94.4	441.0	48.0	1021.1	1.4	642.7	21.0	6.5	0.1
low	6.5	6.0	975	0.2	838.0	221.0	29.0	29.0	115.0	11.4	334.3	2.6	232.6	0.5	0.1	0.05
HAGEL BED																
x̄	9.5	7.1	2264.3	0.4	1774.0	94.6	105.3	74.3	297.4	10.7	681.4	8.0	668.3	7.0	1.3	0.1
s	2.9	0.5	1041.8	0.5	1056.3	---	116.7	72.3	202.3	5.5	218.5	9.0	604.8	6.3	2.4	0.2
n	20	20	20	13	19	1	17	17	17	17	18	3	19	19	14	14
high	14.5	8.0	3750	1.9	3874.0	---	337.0	202.0	810.0	23.2	1098.0	14.3	2087.0	28.0	8.5	0.7
low	3.0	6.4	900	0.1	649.0	---	8.0	2.1	72.4	4.2	412.2	1.6	41.2	0.5	0.0	0.0
SAND 5m BELOW HAGEL BED																
x̄	9.2	7.4	3453.8	0.6	2644.8	---	153.2	80.4	567.7	10.2	835.0	---	1217.0	6.3	0.6	0.4
s	2.0	0.9	932.4	0.5	985.0	---	136.1	74.8	247.6	2.2	327.7	---	663.3	3.3	1.0	0.5
n	13	13	13	13	10	---	13	13	13	13	13	---	13	13	13	13
high	12.0	9.3	4400	1.6	4428.0	---	431.8	260.0	860.2	14.0	1288.2	---	2539.0	11.0	2.9	1.4
low	5.5	6.4	1900	0.0	1413.0	---	30.9	12.7	5.9	.74	106.8	---	415.0	2.0	0.0	0.0
SILT 15m BELOW HAGEL BED																
x̄	8.7	7.6	2946.5	0.5	1983.2	78.6	53.6	35.7	636.5	10.5	1154.7	12.8	694.7	11.1	2.2	0.7
s	2.6	0.5	1345.0	0.5	911.0	40.2	52.0	41.9	252.1	6.8	413.6	0.3	658.0	15.6	8.7	0.7
n	47	47	47	35	44	4	47	47	47	47	47	13	47	47	41	38
high	14.0	8.4	8000	2.7	5056.0	128.0	191.5	186.0	1311.0	36.4	1942.2	13.0	1642.6	107.2	55.8	2.3
low	3.0	5.9	600	0.1	416.0	43.3	3.5	2.8	106.5	3.5	248.0	12.6	140.9	0.5	0.0	0.0

TABLE 2 (Continued)

	SO₃	B	Alk	Fe	Mn	Cd	Hg	Se	As	Pb	Ba	Cr	Co
		mg/L						µg/L					
SPOILS													
x̄	3.8	1.0	954.7	555.6	1951.2	1.9	0.3-	1.6	3.7	4.9	92.3	7.3	13.3
s	1.8	1.2	317.0	557.7	1100.4	2.4	0.3	1.2	2.9	7.2	93.5	4.9	11.9
n	14	11	15	16	16	16	15	12	12	10	10	10	16
high	8.0	4.5	1732.0	1470.0	4070.0	8.8	1.3	4.8	10.1	22.1	348.5	18.1	46.1
low	1.5	0.5	298.0	20.0	2.8	0.5	0.0	0.0	0.4	0.0	10.4	1.9	1.8
KINNEMAN CREEK BED													
x̄	5.0	---	507.0	484.0	325.8	12.0	0.1	2.5	1.0	7.1	51.6	2.5	6.6
s	4.2	---	237.8	369.4	200.3	16.7	0.2	1.3	---	5.4	12.0	1.6	1.1
n	7	---	4	8	6	6	2	6	1	5	5	2	2
high	8.0	---	837.0	1300.0	610.0	34.0	0.2	3.4	---	13.8	60.0	3.6	7.4
low	2.0	---	274.0	45.1	95.0	0.2	0.0	1.6	---	2.0	43.1	1.4	5.8
HAGEL BED													
x̄	4.9	0.9	655.2	220.2	1072.7	1.2	0.4	0.8	2.8	16.9	159.6	6.0	8.9
s	6.1	0.9	399.2	156.4	1120.9	1.1	0.2	1.1	2.1	25.5	257.3	6.7	9.5
n	9	6	16	16	15	15	2	9	6	8	8	8	14
high	20.5	2.8	2040.0	580.0	3440.0	4.4	0.7	2.9	6.5	76.5	792.0	20.0	33.5
low	0.5	0.5	338.5	50.0	50.0	0.5	0.2	0.0	1.0	0.4	42.7	0.3	0.0
SAND 5m BELOW HAGEL BED													
x̄	5.9	3.5	684.5	236.2	660.0	5.0	0.3	1.6	8.1	11.5	54.2	13.0	21.6
s	8.6	2.3	268.7	222.9	765.6	7.1	0.3	1.8	13.4	12.2	22.7	11.4	18.7
n	13	9	13	16	13	13	13	9	9	7	7	7	13
high	31.0	7.2	1055.9	630.0	2930.0	23.5	0.8	6.1	37.6	38.8	79.5	36.8	66.1
low	1.0	0.5	87.5	20.0	120.0	0.5	0.0	0.0	0.7	3.2	7.2	2.9	0.0
SILT 15m BELOW HAGEL BED													
x̄	8.4	5.1	956.9	227.3	265.1	2.6	0.6	2.2	3.6	14.1	154.4	12.9	20.8
s	11.6	10.3	337.4	150.4	144.2	5.2	0.2	1.8	3.4	10.8	197.9	10.5	15.6
n	36	18	39	47	45	45	41	28	27	29	25	25	41
high	55.0	39.3	1592.0	640.0	770.0	25.2	4.6	14.0	14.0	39.0	758.0	52.6	83.9
low	1.0	0.5	203.0	30.0	49.0	0.2	0.0	0.0	0.0	0.9	31.6	2.1	0.2

concentrations in undisturbed sediments and in spoils without fly ash range from 10 to 50 µg/l.

The data in Table 2 indicate that the water in the base of the spoils is generally more highly mineralized, with respect to major ions, than groundwater in undisturbed units. Total-dissolved-solids concentrations in spoils at the Center Mine range from 1599 to 5388 mg/l. Sulfate concentrations range from 565 to 2520 mg/l. However, cadmium, arsenic, and selenium in spoils groundwater are within the same ranges of concentrations as in undisturbed units. All are well below the maximum permissible concentrations for drinking water.

Groundwater in Fly Ash Disposal Areas

Two fly ash disposal areas were instrumented at the Center Mine. In one area, fly ash had been placed in the vee-notch positions between spoil ridges. Monitoring of this site began within six months of closure in 1980. Fly ash in the second disposal site was emplaced in a pit-bottom setting and had been in place for approximately five to six years before monitoring began in 1978.

All groundwater monitoring data from the fly ash disposal sites at the Center Mine can be found in Groenewold, et al. [8]. The leachate generated by the fly ash is typically very alkaline and has extremely high Na, SO_4, and total-dissolved-solids concentrations. Sodium concentrations in fly ash-affected groundwater have been observed as high as 16,947 mg/l and sulfate concentrations have been observed as high as 36,820 mg/l. Total-dissolved-solids concentrations have been observed as high as 52,650 mg/l [8].

Field pH values range from 5.6 to 12.2. The pH values in fly ash leachate-affected groundwater at the Center Mine showed a gradual decrease from 1978 through 1983; analyses obtained during 1984 indicate a rise in pH values (Figure 2).

Soon after disposal in 1980, extremely mineralized groundwater was observed immediately below fly ash emplaced in the unsaturated vee-notch setting. These concentrations later showed a significant decrease through 1984. Drilling to obtain cores of the fly ash in this setting has indicated that the inherent pozzolanic characteristics of the fly ash have produced a relatively high-strength material. Coring of the fly ash became impossible two years after disposal. These data suggest an initial dissolution and flushing of contaminants from the vee-notch fly ash site. Because the fly ash in these settings is above the water table and the infiltration rate is very low [2], this initial flushing is likely due to water entrapped in the vees during disposal and burial operations.

Analyses of water from pressure-vacuum lysimeters emplaced below the vee-notch disposal sites and obtained soon after closure indicate high concentrations of major ions, particularly Na and SO_4, which exceed those of groundwater from piezometers screened in the underlying saturated zone. The extreme mineralization of both the porewater and groundwater in the vee-notch disposal areas have shown a continuous decrease since the initial water samples were obtained.

In contrast to the vee-notch setting, the concentrations of the major species in groundwater adjacent to a saturated pit-bottom disposal setting at the Center Mine have remained relatively constant since first sampled in 1978. As previously noted, this site was not instrumented until several years after emplacement of the fly ash in 1973 and 1974. Thus, these data are most useful for evaluating long-term leaching characteristics of fly ash

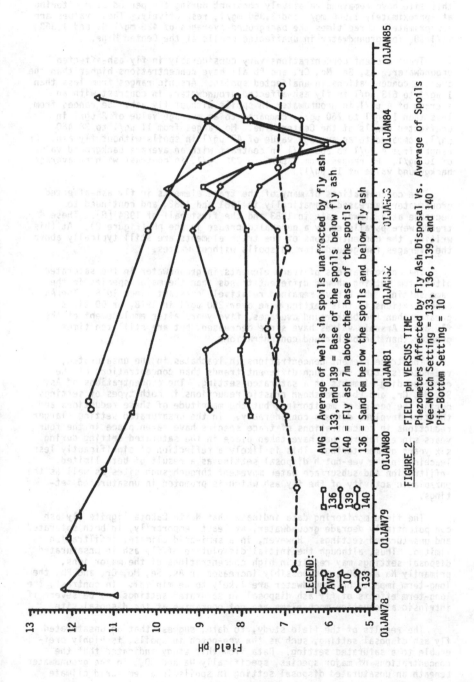

LEGEND:

AVG — ▽
10 — △
133 — ◁

136 — ◇
139 — ○
140 — □

AVG = Average of wells in spoils unaffected by fly ash
10, 133, and 139 = Base of the spoils below fly ash
140 = Fly ash 7m above the base of the spoils
136 = Sand 6m below the spoils and below fly ash

FIGURE 2 - FIELD pH VERSUS TIME
Piezometers Affected by Fly Ash Disposal Vs. Average of Spoils
Vee-Notch Setting = 133, 136, 139, and 140
Pit-Bottom Setting = 10

in this type of setting. The Na and SO₄ concentrations in groundwater at this site have remained relatively constant during the period of monitoring at approximately 1,500 mg/l and 3,350 mg/l, respectively. These values are approximately three times the background averages of 426 mg/l Na and 1,363 mg/l SO₄ for groundwater in unaffected spoils at the Center Mine.

Trace element concentrations vary considerably in fly ash-affected groundwater. As, Se, Mo, Cr, and Pb all have concentrations higher than the average concentrations in unaffected spoils. Arsenic ranges from less than 1 μg/l to 613 μg/l in fly ash-affected groundwater, in contrast with an average of 4 μg/l in groundwater in spoils without fly ash. Se ranges from less than 1 μg/l to 760 μg/l, compared to an average value of 2 μg/l in unaffected spoils at the Center Mine. Mo ranges from 11 μg/l to 38,460 μg/l, compared to an average value of 47 μg/l in spoils without fly ash. Cr ranges from 4 μg/l to 205 μg/l in contrast with an average background value of 10 μg/l. Pb ranges from 2 μg/l to 236 μg/l, in contrast with an average background value of 15 μg/l.

The concentrations of many of the trace elements in fly ash-affected groundwater decreased dramatically in 1981 and 1982, and continued to decrease at a slower rate in 1983 and the first half of 1984 [8]. These trends were paralleled by a general decrease in the pH (Figure 2). At this writing, the concentrations of the trace elements are still typically above the averages for groundwater in spoils without wastes.

The concentrations of trace elements in groundwater in the saturated pit-bottom setting showed different trends than the major species in the same setting, which have remained relatively constant since 1978. The As concentrations in this setting were over 200 μg/l in 1978, or 50 times greater than the background averages, five years after emplacement of the fly ash. Arsenic values have since decreased, but are still ten times greater than the background concentrations.

The trace element concentrations in leachates in the unsaturated, vee-notch setting show much different trends than concentrations of the corresponding elements in a saturated setting. The concentrations of As, Se, Mo, Cr, and Pb have shown drastic reductions in both types of settings during the period of monitoring, but the magnitude of these reductions and the time involved differs considerably. In the unsaturated setting, larger reductions in concentrations of trace species have taken place in the four years since closure than have taken place in the saturated setting during six years of monitoring. This is likely a reflection of significantly less leaching in the vee-notch disposal setting as a result of both limited infiltration and subsurface water movement through such sites as well as the pozzolanic activity of the fly ash which is promoted in unsaturated settings.

The field monitoring data indicate that North Dakota lignite fly ash can potentially degrade groundwater, at least temporarily, in both saturated and unsaturated settings. However, in a semi-arid climate, infiltration is limited. Thus, although the initial dissolution of fly ash in unsaturated disposal settings may result in high concentrations of the major ions, primarily Na and SO₄, and possibly increases in As, Se, Mo, Cr, and Pb, the long-term impacts on groundwater are likely to be minimal. In contrast, the long-term effects of fly ash disposal in saturated settings may be severe if intrinsic attenuation mechanisms are not operative at the disposal site.

The results of the field study, to date, suggest that an unsaturated fly ash disposal setting, such as the vee-notch in spoils, is highly preferable to a saturated setting. Data from this study indicated that the concentrations of major species, specifically Na and SO₄, in the groundwater beneath an unsaturated disposal setting in spoils in a semi-arid climate will not be significantly greater than values for unaffected spoils after two to three years. Trace species in such a setting will also likely show similar decreases. A saturated setting such as a pit bottom, in contrast, will likely be characterized by highly mineralized groundwater in areas adjacent to the fly ash for many years after disposal.

Alkaline Neutralization Studies

Over 300 different experiments were performed. Three sets of neutralization capacity measurements were made. Table 3 summarizes the results of these experiments. In each set, at least two replications of each measurement were run. In Table 3, Set I corresponds to the technique in which all samples were run at a given concentration of base before continuing on to the next concentration of base. Set II corresponds to the technique in which a given sample was run simultaneously in three concentrations of base. Set III corresponds to data taken in the same manner as Set I, but $Ca(OH)_2$ and additional NaOH concentrations were used. Also, Set III was run in Teflon labware instead of glass. The concentrations of the 0.1, 0.5 and 1.0 M NaOH are correct to the single significant figure quoted, but were actually standardized to three significant figures. At that level of precision the numbers would be slightly smaller than the values presented. As may be deduced from Table 3, neutralization capacities between different samples and different techniques are quite reproducible at lower base concentrations, but become less so at high base concentrations.

Two clear trends are evident from Table 3. 1) Neutralization capacity of all soils is dependent upon the concentration of base being neutralized, and 2) at equivalent base normalities, neutralization capacity when Ca^{++} is the counter ion is approximately double the neutralization capacity when Na^+ is the counter ion. There appears to be little dependence of neutralization capacity among the eight soil types tested.

Alkaline extracts obtained in experiment Set III were analyzed for Si and Al. These extracts were compared to extracts prepared by shaking the samples in 400 ml of distilled water. With one exception (Al in oxidized clay), soluble Si and Al concentrations were very small in distilled water. Soluble Si and Al were also very small in alkaline solutions of $Ca(OH)_2$. In NaOH, however, there appears to be a relationship between base concentration and soluble Si and Al. With the exception of sample #6 where the data are limited, soluble Al increased in every case as the concentration of NaOH increases. The trend is less regular for soluble Si, but more Si is dissolved at higher base concentrations that at lower.

Trace Element Attenuation Studies

Trace element attenuation experiments have indicated the following:

1. Arsenic introduced as As^{+5} is highly attenuated (50 - 95% removed). It was found that as alkali was added to the systems studied, arsenic adsorption decreased.

2. Selenium introduced as Se^{+4} is attenuated to a lesser extent than arsenic (30 - 80% removed). As alkali was added to the systems studied, selenium adsorption decreased.

3. Molybdenum introduced as Mo^{+6} appeared to remain virtually unattenuated. From a practical standpoint, this was unfortunate but not entirely unexpected. As described earlier, test piezometers below disposal sites at the Center Mine have shown high, but transient, Mo levels. This probably indicates that highly mobile unattenuated Mo species were traveling with water flowing through this disposal area [9].

4. Cadmium is removed in excess of 99 percent. This was to be expected in a system with a pH in excess of 6.5 - 7.0. At this pH and above, heavy metals such as cadmium and lead tend to precipitate as hydroxide carbonates [10]. In a system with a low pH, one would normally expect cadmium to behave much like calcium since the ratios of their adsorption selectivity coefficients is about 1. The exception to this is with the hydrolyzed ion $CdOH^{-2}$. This species exhibits increased competition over calcium.

These data show a strong correlation with the pH and trace element trends in fly ash-affected groundwater at the Center Mine study sites. In addition, these results clearly reinforce our conclusions regarding the pH neutralization capacity of western North Dakota overburden sediments.

Field and laboratory studies both indicate a relationship between alkalinity and mobility of trace elements, particularly arsenic and selenium [11]. Thus, any mechanism within disposal site sediments that either buffers or neutralizes the alkaline fly ash leachate will significantly decrease the mobility of the toxic species. It is readily apparent from Table 3 that all of the North Dakota overburden materials tested have the capacity to neutralize alkaline solutions to some degree. For a solution of NaOH of pH = 13, one may expect that one gram of overburden material will neutralize 0.3 - 0.4 millimoles of the base. Stated another way, if NaOH may be considered a nominal base, these overburden materials may be expected to neutralize 1.2 - 1.6 percent of their own weight of that nominal base. In English units, one ton of overburden material could be expected to neutralize 24 - 32 pounds of NaOH dissolved in a 0.1 M (pH = 13) solution. Higher pH values are unlikely to exist in field situations and lower pH values imply less alkaline material to be neutralized.

In principle, if one had an estimate of the total extractable alkalinity of a given mass of fly ash, one could use the values obtained above to estimate the possible penetration of alkaline leachate into the materials surrounding the disposal site. The most serious errors would probably arise from assumptions that would have to be made concerning the intimacy of contact between the migrating solution and the soil materials. The trend of decreasing groundwater pH values during the period of monitoring of fly ash disposal sites at the Center Mine suggests that indeed, alkaline neutralization of the leachate is occurring. The groundwater pH in specific settings, particularly those in close proximity to the disposal sites, can be expected to rise once the neutralization capacity of the sediments along the flow path between the fly ash and the specific monitoring point has been exceeded.

While the mechanisms of neutralization are necessarily speculative, there is no doubt that under the conditions of our experiments significant neutralization of base occurs at all concentrations above background levels. Even more striking is that significant neutralization occurs for all of the varieties of soil materials tested, from reduced silts to oxidized sands. In light of this work, the requirement of liners in fly ash disposal sites must be carefully evaluated. If liners are required to retard the movement of major ions, e.g. sulfate, our work is not relevant. On the other hand, if liners are being installed specifically to retard the movement of trace elements, then these data might reduce the justification for such liners. For west-central North Dakota, at least, one soil material appears as effective as another in moderating leachate pH and thus, presumably, trace element mobility.

Continued studies focused on the definition of attenuation mechanisms are essential to a realistic evaluation of the long-term impact of coal-processing wastes on groundwater. In turn, long-term field testing and

TABLE 3

NEUTRALIZATION, Si AND Al IN ALKALINE EXTRACTS

Soil Sample	Conc. Base	Set I	Set II	Set III Si(mg/l)	Al(mg/l)
		mmol base neutralized/gram		Si(mg/l)	Al(mg/l)
1	0.0168 Ca(OH)$_2$			0.458	6.6	--
	0.035 NaOH			0.235	160	22
	0.1 NaOH	0.389		0.346	207	18
	0.5 NaOH	0.957	0.870	1.100	250	55
	1 NaOH	2.200	1.380	1.100	242	109
	H$_2$O				--	--
2	0.0168 Ca(OH)$_2$			0.384	5.5	9.8
	0.035 NaOH			0.170	78	39
	0.1 NaOH	0.307	0.332	0.478	110	6.2
	0.5 NaOH	0.820		0.884	214	110
	1 NaOH	1.920	1.490	1.340	508	204
	H$_2$O				--	243
3	0.0168 Ca(OH)$_2$			0.314	3.4	1.3
	0.035 NaOH			0.162	37	8.3
	0.1 NaOH	0.339	0.360	0.295	48	32
	0.5 NaOH	1.040	0.822	0.627	2240	75
	1 NaOH	1.580	1.270	0.706	1230	130
	H$_2$O				--	0.002
4	0.0168 Ca(OH)$_2$			0.383	0.5	--
	0.035 NaOH			0.112	35	15
	0.1 NaOH	0.432	0.322	0.199	40.2	22
	0.5 NaOH	0.858	0.840	0.805	240	116
	1 NaOH	1.920	1.400	1.120	120	236
	H$_2$O				--	0.9
5	0.0168 Ca(OH)$_2$			0.368	5.7	10
	0.035 NaOH			0.225	58.7	47
	0.1 NaOH	0.338	0.315	0.337	45.9	24
	0.5 NaOH	0.975	0.940	0.552	185.5	173
	1	1.460	0.920	1.200	176.3	196
	H$_2$O				--	1
6	0.0168 Ca(OH)$_2$				--	--
	0.035 NaOH			0.185	60	185
	0.1 NaOH	0.276	0.220		--	--
	0.5 NaOH	0.759			--	--
	1 NaOH		0.870	1.360	217.4	109
	H$_2$O				--	0.24
7	0.0168 Ca(OH)$_2$			0.469	11.4	--
	0.035 NaOH			0.185	255	24
	0.1 NaOH	0.414	0.378	0.381	316	24
	0.5 NaOH	0.957	0.889	1.360	688	42
	1 NaOH	1.740	1.230	1.330	561	95
	H$_2$O				--	0.4
8	0.0168 Ca(OH)$_2$			0.473	8.5	--
	0.035 NaOH			0.248	255	10
	0.1 NaOH	0.394		0.307	316	16
	0.5 NaOH	0.879	0.799	0.900	688	45
	1 NaOH	1.560	1.100	1.520	561	82
	H$_2$O				--	0.2

monitoring will be required to determine the relevance of laboratory results to field settings.

REFERENCES

1. Groenewold, G.H., Cherry, J.A., Manz, O.E., Gullicks, H.A, Hassett, D.J., Rehm, B.W., 1980, "Potential Effects on Groundwater of Fly Ash and FGD Waste Disposal in Lignite Surface-Mine Pits in North Dakota," Symposium on Flue Gas Desulfurization, Houston, Texas, October 28-31, pp. 657-693.
2. Groenewold, G.H., Koob, R.D., McCarthy, G., Rehm, B.W., Peterson, W., 1983, "Geological and Geochemical Controls on the Chemical Evolution of Subsurface Water in Undisturbed and Surface Mined Landscapes in Western North Dakota," North Dakota Geological Survey Report of Investigation No. 79, 299 p.
3. U.S. Environmental Protection Agency, 1977, Procedures Manual for Groundwater Monitoring at Solid Waste Disposal Facilities: Cincinnati, Ohio, 269 p.
4. Groenewold, G.H., Hemish, L.A., Cherry, J.A., Rehm, B.W., Meyer, G.N., Clayton, L.S., Winczewski, L.M., 1979, "Geology and Geohydrology of the Knife River Basin and Adjacent Areas of West-Central North Dakota," North Dakota Geological Survey Report of Investigation No. 64, 402 pp.
5. Griffin, R.A., Roy, W.R., Chow, S.F.J., 1982, Batch Sorption Procedures for Solute Migration Study, Illinois State Geological Survey, 7 p.
6. Karimian, N., Cox, F.R., 1978, Adsorption and Extractability of Molybdenum in Relation to Some Chemical Properties of Soil, Soil Sci. Soc. Am. J., V. 42, pp. 757-761.
7. Garcia Mirogaya, J., Page, A.L., 1976, Influence of Ionic Strength and Inorganic Complex Formation on the Sorption of Trace Amounts of Cd by Montmorillonite, Soil Sci. Soc. Am. J., V. 40, pp. 658-663.
8. Groenewold, G.H., Stadum, M.A., Koob, R.D., Manz, O.E., Hassett, D.J., Hassett, D.F., 1984, "Disposal of Fly Ash and Fly Ash Flue Gas Desulfurization Waste in a Western Decoaled Strip Mine," Interim Report to DOE, 150 p.
9. Jones, L.H.P., 1957, The Solubility of Molybdenum in Simplified Systems and Aqueous Soil Suspensions, Journal of Soil Sci., V. 8, No. 2, pp. 314-327.
10. Lindsay, W.L., 1979, Chemical Equilibria in Soils, John Witey and Sons, New York, NY, 449 p.
11. Griffin, R.A., Fiort, R.R., 1977, Effect of pH on Adsorption of Arsenic and Selenium from Landfill Leachate by Clay Minerals, Soil Sci. Soc. Am. J., V. 41, pp. 53-57.

LEGAL NOTICE

Work supported by DOE Contract DE-AB18-80FC10120 and by the Gas Research Institute Contract No. 5082-253-0771 through the North Dakota Mining and Mineral Resources Research Institute.

MOBILITY OF ORGANIC AND INORGANIC CONSTITUENTS FROM ENERGY AND COMBUSTION-RELATED WASTES UNDER CODISPOSAL CONDITIONS

M. P. MASKARINEC[a], C. W. FRANCIS[b], J. C. GOYERT[c]
[a]Analytical Chemistry Division and [b]Environmental Sciences Division, Oak Ridge National Laboratory, P. O. Box X, Oak Ridge, TN 37831
[c]Science Applications, Inc., Oak Ridge, TN 37831

(Received 29 November, 1984; accepted 20 February, 1985; Refereed)

ABSTRACT

The criteria for determining toxicity of a solid waste under the Resource Conservation and Recovery Act (RCRA) are the levels of NIPDW contaminants as determined using the extraction procedure (EPA-EP). This procedure is predicated on a scenario of disposal of the solid waste in a municipal landfill, resulting in contact with an acidic leaching medium. The scenario includes a 95/5 municipal waste/industrial waste ratio. The experimental definition of the leaching rates of various contaminants under the conditions of this scenario is the object of this work.

The experimental design included the use of large scale lysimeters (1.8 M in diameter x 3.6 M height) packed with 1.5 metric tons of municipal waste. A flow rate of -6.5 l/day of distilled water was added to the lysimeters for the generation of municipal waste leachate (MWL). The MWL was used to leach a variety of industrial wastes, including resource recovery ashes, an API separator sludge/incinerator ash mixture, and a coking plant wastewater treatment sludge. The resulting leachates were analyzed for both inorganic and organic constituents.

Inorganic constituents showed two distinct types of leaching behavior. Freely soluble elements (B, Na, Ca, etc.) gave leaching curves which showed exponential declines. Acid-soluble elements (Ni, Zn) eluted as "peaks" as the pH of the leachate decreased. Organic constituents gave leaching curves which were dependent on solubility: either an exponential decline (freely soluble compounds) or a relatively flat curve at or near the solubility limit. The availability of this data base has allowed the development of a laboratory extraction procedure for the prediction of contaminant migration in municipal landfills.

INTRODUCTION

To assess the possibility of groundwater contamination from the disposal of coal conversion and combustion wastes in nonsecure landfills, laboratory extraction procedures have been developed. The best known of these is the procedure developed by the USEPA and called the EP [1]. This procedure is based on the premise of leaching in a municipal landfill (therefore at low pH) with a ratio of 95% municipal refuse/5% industrial waste. Several other laboratory extraction tests are also available, although the differences between tests may not be great in all cases. All of these procedures were developed based on predictions of leachability in field situations and without a great deal of field data. The objective of this work was to develop a data base from which values for contaminant concentrations could be extrapolated, and then to use this data base for comparison with concentrations obtained with laboratory extraction procedures. The scenario chosen

for study was the codisposal scenario of the EP. The strategy employed
involved 1) the generation of municipal waste leachate (MWL) in large
lysimeters, and 2) the use of the MWL to leach coal conversion and
combustion wastes under simulated landfill conditions.

MATERIALS & METHODS

Approximately 1.5 metric tons (Mg) of municipal refuse from the
city of Oak Ridge, TN were placed in each of four large lysimeters
(lined concrete cylinders 1.8 m diameter and 3.6 m high). The municipal
waste consisted of a residential waste (collected primarily from house-
holds) and a commercial waste (collected primarily from the downtown
shopping area and fast-food restaurants). Each lysimeter was filled
as shown in Figure 1. The flanged drain was packed with glass wool
and covered with pea gravel to allow for adequate flow while removing
some of the particulate matter. No attempt was made to simulate the
packing density of a commercial landfill, although the density of
the material in the lysimeters was approximately 280 kg/m^3, about
half the density of a commercial facility [2]. The clay loam soil
added to the top of the lysimeter prevented moisture and vapor loss.
Distilled water was added to the top of the lysimeters at a rate of
26.5 L/day. Hydraulic residence time for each of the lysimeters was
an estimated 30 to 40 days. Leachate from each of the lysimeters
was collected in 30 L glass carboys which served as sumps for recycling
pumps (10 to 15 ml/min). Glass manifolds were used to divert MWL
to the solid waste columns and to provide access for the electrodes
necessary to measure pH and Eh (a combination glass pH and Ag/AgCl
reference electrode and a Pt working electrode). Eh is measured using
these electrodes and provides an estimate of the reducing or oxidizing
capacity of the leachate. Therefore, samples collected for laboratory
analysis of the MWL were taken at the sump while the real-time characterization was done in the manifold.

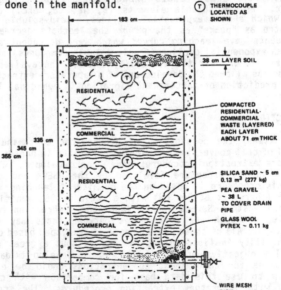

Figure 1. Schematic of Municipal Waste Lysimeter

The MWL was diverted to industrial solid waste columns at a flow rate of 0.8 ml/min using peristaltic pumps. The industrial waste columns were made of borosilicate glass 38.7 cm i.d. x 30.5 cm high. The columns were covered with a plexiglass plate (2 cm thick), whose underside was covered with an adhesive-backed Teflon overlay. The waste itself was contained between two 5 to 7 cm layers of sand. Between the waste and the top layer of sand was a thin layer of glass wool used to further facilitate the lateral dispersion of the MWL. The quantity of waste in each column was 3.6 kg. Leachate from the industrial waste columns was collected in Tedlar (PVF) gas sampling bags (SKC Inc.,Eighty Four, Pa.) [3]. The experiments were carried out in two phases, each employing two lysimeters. The first phase consisted of 79 days of leaching and the second phase 103 days. The objective was to assure that the leaching continued to a liquid/solid ratio of at least 20.

The industrial solid wastes represented a variety of waste types and sources. Only those related to coal conversion and combustion will be considered in this discussion. The wastes were chosen on the basis of expected high concentrations of various contaminants in the leachates. Two wastes, an admixture of a petroleum refining incinerator ash and an American Petroleum Institute (API) separator sludge (an oily liquid) and the ammonia lime still bottoms from a coking plant (a high - Ca, water treatment plant sludge), contained both organic and inorganic constituents which were leachable. Four additional wastes consisted of resource recovery ashes which contained only inorganic constituents at levels above the MWL control. These wastes resembled flyash physically, but had been formed at lower temperature and had a wider particle size distribution. The resource recovery ashes were also leached with distilled water to simulate a monofill situation.

Laboratory extractions were conducted using a rotary-batch extractor at a liquid/solid ratio of 20/1. An extractor equipped with six two liter borosilicate glass vessels was rotated at 30 rpm over an 18 hour period (overnight). The extraction media were (1) pH 5 sodium acetate (0.1 M with respect to acetate; this is equivalent to the maximum acetate added in the EP), and (2) a weak carbonic acid solution (carbon dioxide saturated deionized distilled water). In addition, EP extractions [1] were carried out on all wastes for comparison purposes.

Inorganic analysis was carried out as follows. All wastes were analyzed directly by Neutron Activation Analysis (NAA) and, after digestion, by Inductively Coupled Plasma Spectrometry (ICP). The two independent measurements were used to obtain accurate estimates of the total mass of inorganic materials in each waste. These methods were previously used to characterize the inorganic content of other wastes [4]. The ICP technique was also used for multielement analysis of both leachates from the lysimeter and extracts from the laboratory procedures. In the case of the MWL, the calibration standards were compared to spiked solutions of MWL in order to assure that the interelement correction factors used in the analysis were correct. Certain selected extracts were analyzed for inorganic constituents using Atomic Absorption Spectrophotometry (AAS) using the technique of standard addition.

Organic analyses of the API separator sludge/incinerator ash and ammonia lime still bottoms were carried out in two phases. First, the waste was extracted with methylene chloride and analyzed by gas chromatography (GC) and GC/Mass Spectrometry. Those contaminants which were found at concentrations sufficient for consideration as target analytes were then subjected to quantitative analysis by methods similar to those employed in the leachate and extract analysis. Analysis of the API separator sludge/incinerator ash admixture for PAH was carried out by HPLC/fluorescence detection. Analysis of the ammonia lime still bottoms for polynuclear aromatic hydrocarbons (PAH), was carried out by GC with flame ionization detection.

RESULTS AND DISCUSSION

Leachate from the lysimeters was monitored relative to pH, Eh, organic acids, and inorganic constituents during the times in which MWL was used as a leaching medium. Values of pH ranged from 3.5 to 6.2. Carboxylic acid concentrations ranged from 3000 to 2000 mg/L in the first experiment and from 4000 to 3000 mg/L in the second experiment. The acids accounted for 60 to 90 % of the total organic carbon content of the leachate. The concentration of monovalent cations decreased over the leaching period. For example, sodium concentration decreased from 600 to 115 mg/L and from 340 to 150 mg/L respectively in the two experiments. Most of this reduction occurred during the first 30 days of leaching. Other than the acids, the dominant anion was chloride. The concentration of chloride also decreased primarily during the first 30 days of leaching, remaining relatively constant thereafter. Concentrations of organic and inorganic constituents in the MWL did not differ appreciably from lysimeter to lysimeter.

Concentrations of organic and inorganic constituents in the industrial wastes are given in Table 1. While these are not the only constituents in the wastes, they are the only constituents which were not found in the MWL background. As stated earlier, the objective of the lysimeter experiment was to define the time course of contaminant mobility in general and to generate target concentrations of the individual contaminants for comparison with the laboratory extractions.

Numerous factors contribute to the leaching pattern of the various inorganic elements present in the waste leachates. Leaching is typified by two basic mobility curves: (1) a curve that depicts the rapid leaching of readily soluble constituents from the wastes, and (2) a curve in which the initial release is slow, with the concentration of the contaminant increasing with time. The first type of curve is exemplified by the leaching of calcium from the resource recovery ash 2 (Figure 2). The wastes reported on here consistently showed this type of behavior. The second type of curve has been noted in other wastes [5]. Complete data for the leaching of inorganic compounds from the resource recovery ashes have been published elsewhere [6].

The leaching behavior of the organic compounds followed similar trends. The very soluble phenols in the ammonia lime still bottoms leached at a rapid rate, as indicated by Figure 3. Essentially all of the hydroxybenzenes were leached from this waste at a liquid/solid ratio of 4. It should be noted that this waste had a high pH, such that these compounds existed in ionic form, further enhancing solubility. In the case of the neutral, less soluble PAH compounds, the rate of mobility was dependent on solubility and partition (Figure 4). The mobility of phenanthrene from the ammonia lime still bottoms is basically independent of time. While the plot appears to be highly variable, it should be noted that although the difference between the high and low value is a factor of ten, the highest value is only 0.2 mg/L. Most of the PAH compounds behaved in a similar fashion. These are

TABLE 1. CHEMICAL CONCENTRATIONS OF WASTES USED IN LYSIMETER STUDY (UG/G)

CHEMICAL	API SEP.SLUDGE/ INCINERATOR ASH	AMMONIA LIME STILL BOTTOMS	RECOVERY ASH 1	RECOVERY ASH 2	RECOVERY ASH 3	RECOVERY ASH 4
Al		5350	53000	60000	54000	45000
Ba		<15	1600	810	1600	1000
Ca	8.9E4	5.5E4	8.5E4	7.9E4	8.8E4	4.2E4
Cr	3900					
Cu			550	2000	650	1700
Fe			86000	58000	36000	63000
K	3400	20800	11000	5800	12000	5500
Mg		17100	16000	7700	8800	6500
Mn			1200	1500	1000	1100
Mo	82		290	170	200	260
Na	9800		18000	9500	18000	7700
Sr	260	235	270	380	300	220
PHENOL		535				
m&p-CRESOL		786				
o-CRESOL		106				
NAPHTHALENE	20	91				
ACENAPHTHENE		0.7				
PHENANTHRENE		84				
FLUORANTHENE		60				
PYRENE		46				

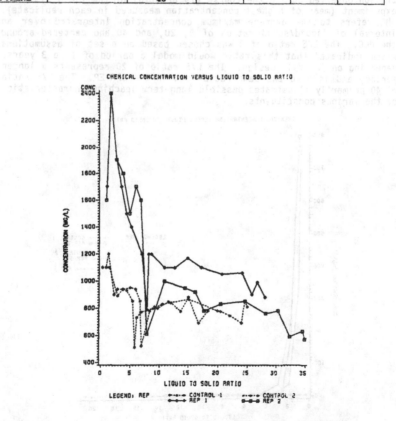

Figure 2. Leaching of Calcium Prior from Resource Recovery Ash 2

relatively important data from the standpoint of disposal of coal conversion wastes, since the two classes of organic compounds expected to be present in these wastes are the phenolic compounds and the PAH.

It is also instructive to consider the fraction of the total mass of each contaminant leached from the wastes over the course of the experiments (Table 2). The range of concentrations was from the control level (for several trace elements from the resource recovery ashes) to quantitative removal (for potassium from resource recovery ash 2 and phenol from the ammonia lime still bottoms). Again, the fraction leached depended primarily on the solubility of the particular contaminant. In the case of the resource recovery ashes, the fraction leached by distilled water was in general lower than that leached by MWL.

Target concentrations are the concentrations of the organic and inorganic constituents that the laboratory extraction procedure is asked to model. The concentration found in a laboratory extract is compared to the target concentration as determined using the lysimeter data. Of course, several options exist for the method to establish the lysimeter target concentration. Those used in this study are illustrated graphically in Figure 5. MLC refers to the maximum concentration observed during the course (79 or 102d) of the field lysimeter experiment (mean of highest concentration measured in each replicate). AMC refers to the average maximum concentration integrated over an interval of liquid/solid ratios of 8, 20, and 40 and centered around the MLC. The L/S ratio of 8 was chosen based on a set of assumptions which indicated that this ratio would model a period of 1 to 3 years, depending on landfill design. The L/S ratio of 20 represents a longer period, and is also the same ratio used in the EPA-EP. The L/S ratio of 40 primarily illustrates possible long-term leaching characteristics of the various constituents.

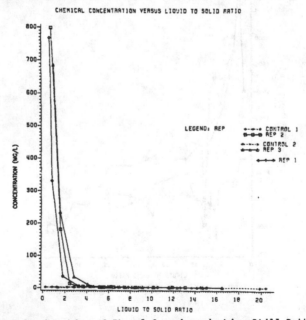

Figure 3. Leaching of Phenol from Ammonia Lime Still Bottoms

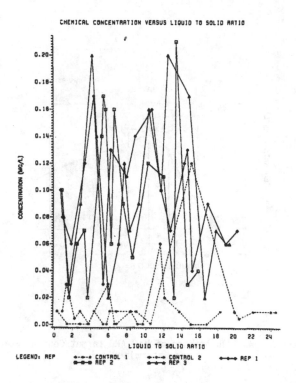

CHEMICAL CONCENTRATION VERSUS LIQUID TO SOLID RATIO

LEGEND: REP   •—•—• CONTROL 1   •—•—• CONTROL 2   •—•—• REP 1
        •—•—• REP 2   •—•—• REP 3

Figure 4.   Leaching of Phenanthrene from Ammonia Lime Still Bottoms

TABLE 2. PERCENT RECOVERY OF TARGET
COMPOUNDS FROM LYSIMETER STUDY (%)

CHEMICAL	API SEP.SLUDGE/ INCINERATOR ASH	AMMONIA LIME STILL BOTTOMS	RECOVERY ASH 1	RECOVERY ASH 2	RECOVERY ASH 3	RECOVERY ASH 4
Al			<0.1	<0.1	<0.1	<0.1
Ba			0.2	9	3	6
Ca	13		9	16	15	12
Cr	17					
Cu			1	1	<0.1	<0.1
Fe			<0.1	3	<0.1	3
K	41		27	97	13	24
Mg			15	8	11	3
Mn			2	<0.1	<0.1	2
Mo	8		<0.1	2	0.1	2
Na	54		13	20	8	9
Sr	11		10	24	12	12
PHENOL		140				
m&p—CRESOL		76				
o—CRESOL		62				
NAPHTHALENE	14	76				
ACENAPHTHENE		67				
PHENANTHRENE		2				
FLUORANTHENE		2				
PYRENE		3				

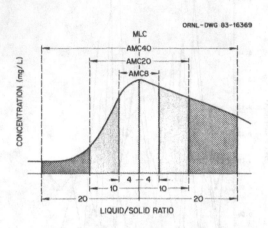

Figure 5. An Illustration of How AMC Target Concentrations
are Determined

In the first phase of these experiments, several extraction procedures
were tested with respect to their ability to approach the target values.
Four media (MWL, 0.1N NaOAc, CO_2/water, and distilled water) were
used at L/S ratios of 2.5, 5, 10, and 20 in both batch and column
modes [5]. The results of this study [5] indicated that the closest
fit with the target values were obtained using CO_2/water and 0.1N
NaOAc in the rotary extractor at a liquid/solid ratio of 20. These
two procedures were then retested in the second phase of the experiment. An example of the comparability of the laboratory procedures
with the field determined MLC and AMC20 values for the organic contaminants
from the ammonia lime still bottoms is shown in Table 3. It is clear
that, at least for this subset of the data, the laboratory procedures
do a respectable job of fitting the field AMC20 values. Of course,
the field MLC values are considerably higher than the concentrations
attained in the laboratory studies. This is due primarily to the
contribution of the high-solubility constituents to the MLC. A thorough
statistical treatment of the data is now in progress, and should result
in a clear understanding of the limitations of laboratory extraction
procedures for the prediction of contaminant mobility in municipal
landfills.

In summary, an experimental regime was established for the generation
of a representative municipal waste leachate. This leachate was then
used to extract contaminants from industrial wastes. The resulting
leaching curves were used to provide target values for the development
of laboratory extraction procedures.

TABLE 3. COMPARISON OF LABORATORY EXTRACTION
PROCEDURES WITH FIELD VALUES (mg/L)

WASTE	CHEMICAL	CO_2	ACETATE	MLC	AMC20
AMMONIA	PHENOL	54.5	36.2	750	36
LIME	M&P CRESOL	28.5	17.2	355	20.6
STILL	O—CRESOL	4.2	3.0	75.6	2.7
BOTTOMS	NAPHTHALENE	2.1	2.0	12.6	3.8
	ACENAPHTHENE	0.03	0.02	0.07	0.02
	PHENANTHRENE	0.15	0.13	0.19	0.07
	FLUORANTHENE	0.05	0.07	0.21	0.07
	PYRENE	0.06	0.06	0.30	0.06
	2—METHYLNAPHTHALENE	0.17	0.13	0.48	0.20
	1—METHYLNAPHTHALENE	0.10	0.07	0.52	0.18
	ACENAPHTHYLENE	0.44	0.32	0.92	0.22

ACKNOWLEDGEMENT

Research sponsored by the U.S. Environmental Protection Agency
under Interagency Agreement DOE No. 40-1017-80, EPA No. 89930007-01-2
under Martin Marietta Energy Systems, Inc. contract DE-AC05-840R21400
with the U.S. Department of Energy.

Although the research described in this report has been funded
wholly or in part by the United States Environmental Protection Agency
(EPA) through an Interagency Agreement to the U.S. Department of Energy,
it does not necessarily reflect the views of EPA and no official endorse-
ment should be inferred.

REFERENCES

1. U.S. Environmental Protection Agency. Identification and Listing
    of Hazardous Waste. In Environmental Protection Agency Hazardous
    Waste Management System. 40 CFR 261.24.

2. Hagerty, D.J., Pavoni, J.L., and Herr, J.E.Jr. Solid Waste Management,
    in Sanitary Landfill, Van Nostrand Reinhold co. 1973, Chapter
    10.

3. Schuetzle, D., Prater, T.J., and Roddell, S.R. Sampling and Analysis
    of Emissions From Stationary Sources I. Odor and Total Hydro-
    carbons. J. Air Pollut. Control Fed. 24(9):925-932, 1975.

4. Griest, W.H., Maskarinec, M.P., and Tomkins, B.A. Chemical and
    Physical Characterization and Management of Feed/Process/Product/-
    Effluent Samples from H-Coal Plant. ORNL/TM-8911, 1983.

5. Francis, C.W., Maskarinec, M.P., and Goyert, J.C. Mobility of
    Toxic Compounds from Hazardous Wastes. ORNL-6044, 1984.

6. Francis, C.W. Leaching Characteristics of Resource Recovery Ash
    in Municipal Waste Landfills. DOE project No. ERD-83-289, Draft
    Final Report, 1984.

INVESTIGATION OF LEACHABILITY OF SUBBITUMINOUS FLY ASH
ENHANCED ROAD BASE MATERIALS

IVAN GARCEZ AND MARTY E. TITTLEBAUM
Department of Civil Engineering, Louisiana State University, Baton Rouge,
LA 70803

(Received 29 November, 1984; accepted 20 February, 1985; Refereed)

ABSTRACT

Enormous amounts of coal fly ash produced by the utilities industry
create a significant disposal problem. The abundance of fly ash along with
its self-hardening properties led the Louisiana Department of Transportation
and Development (LA DOTD) to use fly ash as a soil stabilizer for road bases.
However, the LA DOTD is primarily concerned with the strength of the material
and has not studied its leaching characteristics. In this study, a total of
three samples, a 30 percent subbituminous fly ash/soil mixture, subbituminous fly ash, and soil, were leached following the EPA multiple extraction
procedure and analyzed by Inductively Coupled Argon Plasma Spectrometry
(ICAP). Results indicate that little change in leachate quality of subbituminous fly ash was caused by the soil stabilization process. The
insignificant change is attributed to the low cation exchange capacity of
the soil. ICAP analysis revealed that levels of toxic metals in the
leachates are well within RCRA standards.

INTRODUCTION

The depletion of domestic oil and natural gas reserves, along with the
increasing cost of imported oil, have directed the utilities industry
toward increased use of coal. Such an increase might create a significant
environmental problem because of the generation of three types of wastes:
fly ash, bottom ash, and boiler slag. Among these residues, fly ash poses
the greatest pollution threat because it comprises approximately 80 percent
of the total coal waste and because only a small portion of this massive
production is used commercially [1]. Therefore, if new methods of utilization become economically feasible, they will help reduce the amount of
material that requires disposal.

Because of its self-hardening properties, fly ash has been used in
highway construction, especially as a soil stabilizer for road bases [2,3].
However, the majority of the fly ash stabilization projects implemented by
various highway departments were devoted more to the measurement of
strength and durability of the material rather than to its environmental
hazards [4]. The leaching potential from fly ash structures must be
determined because of the possibility that heavy metals from the ash could
migrate to groundwater systems, contaminating drinking water sources.
Therefore, the quality of leachate from fly ash must be well-evaluated
before using large amounts of it in engineering structures. The specific
objective of this research was to determine the influence of the stabilization process on the leachate quality of subbituminous fly ash when this
pozzolan is used as a road base material.

EXPERIMENTAL METHOD

Materials

Subbituminous fly ash samples were supplied by the Louisiana Department of Transportation and Development (LA DOTD) in Baton Rouge. This ash was collected from the storage silo of Gulf States Utilities, Nelson Station No. 6, in West Lake, Louisiana. The soil used in this investigation was also provided by the LA DOTD. This material, a brown sandy loam soil, came from a pit in Lonepine, Louisiana, during the construction of Interstate Highway I-49.

Acid Digestion

The sample preparation method for determining the bulk analysis of the soil and fly ash was developed by Ritter et al. [5]. This technique involves weighing and transferring a one-gram sample into an acid-washed porcelain crucible. A representative one-gram sample was obtained by quartering according to AASHTO designation T248-74. The sample was ignited at 550°C for 2.5 hours in a muffle furnace. After cooling, the sample was placed in a 30 ml Teflon beaker, covered with 10 ml of concentrated HNO_3 and 10 ml of concentrated HCl and evaporated at 80-90°C to dryness. This procedure was repeated two more times. The residue was dissolved with HCl, then poured into a 50 ml volumetric flask and diluted to volume. This solution was vacuum filtered through a 0.45 μm membrane filter and analyzed in a Jarrell-Ash Inductively Coupled Argon Plasma Spectrometer (ICAP), Model 800, equipped with 15 elemental channels.

Soil Analysis

A chemical analysis of the sandy loam soil was performed at the LSU Agronomy Laboratory. This analysis consisted of the measurement of soil pH and its cation exchange capacity.

Five grams of air-dry soil, greater than 2 mm, was weighed into a 125-ml Erlenmeyer flask and 50 ml of ammonium acetate was added. The flask was swirl, covered and allowed to stand overnight before transferring its contents to a Buchner funnel fitted with a Whatman #42 paper. The soil was leached with an additional 50 ml of ammonium acetate and the leachate transferred to a 100-ml volumetric flask and diluted to volume. This solution was run on a Perkin-Elmer atomic absorption spectrometer for cation determinations and the ammonium saturated soil was used for determination of cation exchange capacity [6].

Leaching Test Protocol

The EPA multiple extraction procedure [7] employed in this investigation to generate leachate from the following samples:

o   Sandy loam soil

o   Subbituminous fly ash

o   Stabilized 30 percent subbituminous fly ash to soil

The term "stabilized" in this particular case signifies molding the specimen by bringing it to its optimum moisture content and compacting it according to standard test ASTM D698-70A. The average 56-day unconfined compressive strength of the mixture was 500 psi.

The multiple extraction procedure consists of a combination of the EP-toxicity procedure followed by successive simulated acid rain extractions. The test protocol listed below was applied for all samples:

(1) A 100-gram sample was sieved through a 100 mesh screen.

(2) The EP-toxicity test was performed on the 100-gram sample as described in "Test Method for Evaluating Solid Waste," USEPA Publication SW 846.

(3) A synthetic acid rain solution was prepared by adding a 60/40 percent mixture of sulfuric and nitric acids to distilled deionized water until the pH was 3.0 ± 0.2.

(4) The solid phase of the sample remaining after step 2 was weighed and placed in a plastic container with 20 times its weight of the synthetic acid rain solution.

(5) The plastic container was then placed in a reciprocating shaker and agitated for 24 hours at room temperature.

(6) The pH of the mixture was recorded at the beginning and end of each extraction.

(7) The aqueous portion was vacuum filtered through a 0.45 μm membrane filter and analyzed in a ICAP, Model 800.

(8) Steps 4-7 were repeated three additional times.

RESULTS

The composition of the subbituminous fly ash sample, Table I, reveals that aluminum, calcium, iron and silicon oxides account for more than 90 percent of the entire ash composition. Concentrations of potassium and

Table I. Bulk Analysis of Sandy Loam Soil and Subbituminous Ash (mg/kg).

Element	Sandy Loam Soil	Subbituminous Ash
Aluminum	9,260.00	13,660.00
Arsenic	79.00	150.00
Cadmium	0.20	1.60
Calcium	119.00	31,270.00
Copper	347.00	220.00
Iron	22,600.00	28,240.00
Lead	112.00	118.00
Magnesium	111.00	181.00
Manganese	127.00	150.00
Molybdenum	12.00	25.00
Nickel	21.00	65.00
Phosphorus	290.00	5,750.00
Potassium	21,860.00	7,290.00
Silicon	478,730.00	225,410.00
Zinc	95.00	153.00

phosphorus are lower but still significant, while the remainder constitutes trace quantities of heavy metals.

The bulk analysis of the sandy loam soil, also listed in Table I, indicates high levels of aluminum, potassium, iron and silicon, followed by minor amounts of the remaining constituents. In addition to bulk chemical analysis, the soil was examined with respect to pH and cation exchange capacity. Results of this analysis are given in Table II.

Tables III through V present the compositions of the various leachates generated in this investigation. Initial and final pH values for each extraction are also included. All elemental concentrations were calculated based on triplicate samples and reported in milligrams per kilogram (mg/kg) of sample. EPA quality control samples for heavy metals were run to check the reliability of the ICAP analysis. Values of this analysis fell within the established EPA confidence intervals.

DISCUSSION

Patterns of Release

Results of this study can best be explained by examining the leachate quality of subbituminous fly ash before and after stabilization with soil.

Table II. Analysis of Sandy Loam Soil Sample.

| pH | meq/100 g | | | | | | CEC at pH 4.7 | CEC at pH 8.1 |
	Organic Matter	P	Ca	Mg	K	Na		
4.7	0.02	0.05	1.6	2.0	0.1	0.2	5.6	9.4

Table III. Leachate Analysis of Subbituminous Fly Ash (mg/kg).

Parameter	Ext. 1	Ext. 2	Ext. 3	Ext. 4	Ext. 5	Sum
Aluminum	322.00	785.00	593.00	503.00	548.00	2,751.00
Arsenic	5.20	6.10	4.40	3.60	4.20	23.50
Cadmium	0.20	N.D.	N.D.	N.D.	N.D.	0.20
Calcium	48,170.00	4,400.00	2,920.00	2,160.00	2,290.00	59,940.00
Copper	0.30	N.D.	N.D.	N.D.	N.D.	0.30
Iron	1.60	2.70	2.60	3.10	2.80	12.80
Lead	2.90	2.40	1.80	1.70	1.90	10.70
Magnesium	11.10	4.00	3.80	4.50	3.80	27.20
Manganese	0.60	1.00	0.80	0.70	0.80	3.90
Molybdenum	4.80	2.00	1.20	0.70	0.70	9.40
Nickel	0.10	N.D.	N.D.	N.D.	N.D.	0.10
Phosphorus	9.30	5.40	4.20	3.70	3.80	26.40
Potassium	103.00	23.10	9.60	9.00	8.10	152.80
Silicon	15.70	28.70	43.10	54.10	54.30	195.90
Zinc	0.20	N.D.	N.D.	N.D.	N.D.	0.20
Initial pH	9.6	9.6	9.6	9.6	9.6	--
Final pH	9.9	10.5	10.5	10.5	10.5	--

N.D. - Non-detectable

Table IV. Leachate Analysis of Sandy Loam Soil (mg/kg).

Parameter	Ext. 1	Ext. 2	Ext. 3	Ext. 4	Ext. 5	Sum
Aluminum	3.40	4.40	4.00	3.40	4.30	19.50
Arsenic	0.20	N.D.	N.D.	N.D.	N.D.	0.20
Cadmium	0.04	N.D.	N.D.	N.D.	N.D.	0.04
Calcium	1.60	2.00	12.00	8.50	9.50	33.60
Copper	0.20	0.10	0.30	3.50	0.10	4.20
Iron	0.60	0.20	0.50	0.30	0.30	1.90
Lead	0.60	0.80	0.70	0.60	0.70	3.40
Magnesium	1.30	0.40	13.10	9.80	10.60	35.20
Manganese	0.10	0.10	1.50	1.10	1.20	4.00
Molybdenum	0.10	0.10	0.20	0.20	0.20	0.80
Nickel	N.D.	0.10	0.10	0.20	0.10	0.50
Phosphorus	0.30	0.30	0.40	0.40	0.30	1.70
Potassium	14.70	16.60	30.70	20.60	14.60	97.20
Silicon	98.60	60.80	45.20	32.80	34.40	271.80
Zinc	0.20	0.10	0.40	0.30	0.30	1.30
Initial pH	4.4	4.4	3.8	3.8	3.8	--
Final pH	4.4	5.3	4.8	4.4	4.4	--

Table V. Leachate Analysis of Stabilized 30 Percent Subbituminous Fly Ash to Sandy Loam Soil (mg/kg).

Parameter	Ext. 1	Ext. 2	Ext. 3	Ext. 4	Ext. 5	Sum
Aluminum	61.80	17.00	11.40	8.60	10.10	108.90
Arsenic	3.70	0.10	0.30	0.30	0.50	4.90
Cadmium	0.90	0.03	N.D.	N.D.	N.D.	0.93
Calcium	30,990.00	1,510.00	700.00	550.00	300.00	34,050.00
Copper	0.50	0.10	N.D.	N.D.	N.D.	0.60
Iron	5.80	3.20	1.00	1.00	1.30	12.30
Lead	2.60	0.50	0.60	0.60	0.90	5.20
Magnesium	5,180.00	325.00	170.00	145.00	74.00	5,894.00
Manganese	14.90	0.70	0.40	0.40	0.10	16.50
Molybdenum	1.40	0.50	0.20	0.10	0.20	2.40
Nickel	3.30	0.20	0.10	0.20	0.20	4.00
Phosphorus	17.10	3.50	2.90	3.40	5.40	32.30
Potassium	123.70	35.80	30.80	27.40	31.40	249.10
Silicon	554.60	341.80	380.40	493.10	251.50	2,021.10
Zinc	0.40	N.D.	N.D.	N.D.	N.D.	0.40
Initial pH	5.0	5.1	5.3	5.2	5.4	--
Final pH	5.0	6.2	6.1	6.0	6.2	--

N.D. - Non-detectable

As Natusch, et al. [8] stated, the relative leachability of elements from fly ash is dependent on the position of the element in the fly ash matrix. The data generated in this investigation revealed that even though constituents such as Al, Fe, K and Si are present in large concentrations within the fly ash structure, very little extraction occurred, particularly in the cases of Al and Fe. Table VI presents the quantity leached of each matrix element as a percent of its bulk concentration. In this table, the fly

Table VI. Percentage Leached of Each Matrix Element from Subbituminous Fly Ash (SFA) Relative to its Original Bulk Concentration (all concentrations reported in mg/kg).

Element	30% SFA/ Soil Leachate	Qty. Leached from Soil	Qty. Leached from SFA	Bulk Concentration of SFA	% Leached
Aluminum	108.90	13.60	95.30	13,660.00	0.70
Calcium	34,050.00	23.50	34,026.50	31,270.00	> 100.00
Iron	12.30	1.30	11.00	28,240.00	0.04
Magnesium	5,894.00	24.60	5,869.40	181.00	> 100.00
Potassium	249.10	68.00	181.10	7,290.00	2.50
Silicon	2,021.10	190.30	1,830.80	225,410.00	0.80

ash/soil mixture was leached and the leachate that came from the 70 percent of soil was subtracted to give the amount of leachate attributed to the fly ash.

The limited leachability of these elements is attributed to the fact that the bulk of the matrix elements is largely concentrated in the interior of the particle. Consequently, their exposure to the aqueous leaching solution is limited. Calcium and magnesium, however, did not follow the same pattern of leaching as shown by their high extractabilities. Their concentrations in the leachates appeared to be higher in magnitude than their bulk concentrations in the fly ash.

Some trace elements present in fly ash are believed to be enriched on the surface layer [9]. If these elements do predominate on the periphery of the particle, the likelihood of being leached out would increase substantially. Table VII gives the quantity leached of each trace metal from subbituminous fly ash after stabilization with soil. Except for Cu and Zn, which were below the detection limits, the remaining constituents exhibited some leaching, especially Cd that appears to be surface-enriched.

Table VII. Percentage Leached of Each Trace Element from Subbituminous Fly Ash (SFA) Relative to its Original Bulk Concentration (all concentrations reported in mg/kg).

Element	30% SFA/ Soil Leachate	Qty. Leached from Soil	Qty. Leached from SFA	Bulk Concentration of SFA	% Leached
Arsenic	4.90	0.10	4.80	150.00	3.20
Cadmium	0.93	0.03	0.90	1.60	56.30
Copper	0.60	0.60	0.00	220.00	0.00
Lead	5.20	2.40	2.80	118.00	2.40
Manganese	16.50	2.80	13.70	150.00	9.10
Molybdenum	2.40	0.60	1.80	25.00	7.20
Nickel	4.00	0.35	3.65	65.00	5.60
Zinc	0.40	0.40	0.00	153.00	0.00

The leached quantities of some metals from subbituminous fly ash and 30 percent subbituminous fly ash/soil mixture were plotted versus the number of extractions. One factor that must be stressed is the difference in pH values between the first and remaining extractions. The acetic acid required to control the pH of the first extraction (EP-toxicity test) to a value of 5.0 ± 0.2, in most cases, provided a favorable environment to mobilize higher quantities of inorganic elements.

Heavy metal release followed an expected pattern of decrease after each extraction. The metal release data showed a variety of trends, such as similar patterns by Al, As and Mo. Plotted in Figure 1(a-c), these elements showed a decrease in concentration with the number of extractions and a reduction in leachability after stabilization with soil. Similarities in release patterns were also displayed by Fe, Mn, Ni, Cd, Cu and Zn. As depicted in Figure 1(d-i), pH appears to have a significant effect in their release. As expected, the non-alkaline leachate yielded the highest extracted quantities of these metals, indicating that they are more mobile in an acid medium. The similar behaviors of Cd, Cu and Zn, which after the first extraction were not detected in the leachate, suggest their possible depletion from the particle surface of fly ash.

Significance of Cation Exchange Capacity

As previously mentioned, the fundamental objective of this study was to examine whether the soil stabilization has any impact on the leachate quality of subbituminous fly ash. A decrease in concentration of inorganic elements released into solution was expected. The basis of this inference is that the soil behaves as a sponge having the ability of exchanging cations, which in turn, are adsorbed onto the surface of its particles. The measurement of this ability is called cation exchange capacity (CEC).

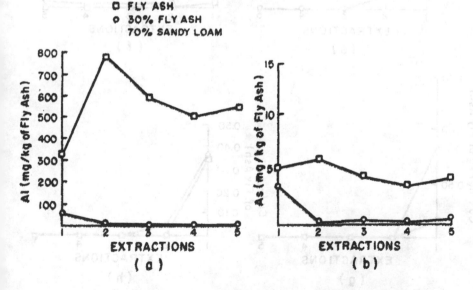

Figure 1. Amount of Various Elements Leached from Subbituminous Fly Ash per Extraction : (a) Al, (b) As, (c) Mo, (d) Fe, (e) Mn, (f) Ni, (g) Cd, (h) Cu, (i) Zn.

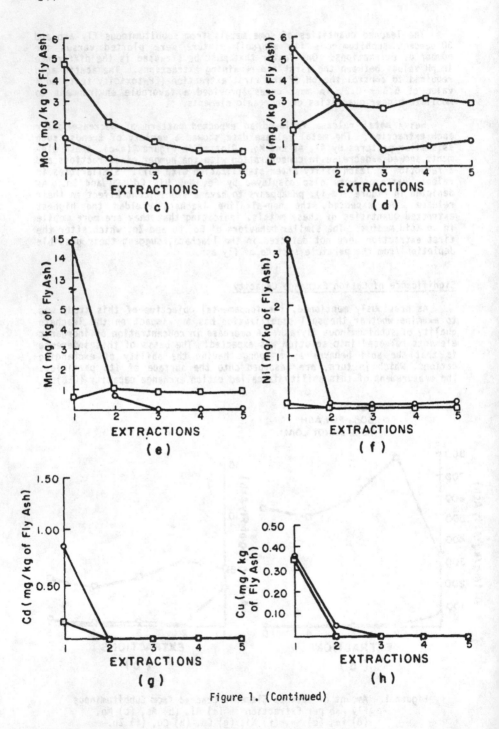

Figure 1. (Continued)

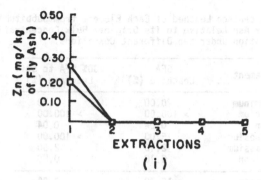

Figure 1. (Continued)

Clay particles may have hydroxyl ions (OH) exposed on their surfaces and edges. At lower pH values, dissociation of OH groups is very weak and the CEC is low. At higher pH values, this capacity increases significantly [10]. The ease with which ions are exchanged depends mainly on the valence and ion size. The greater the charge and smaller the radius of the ion, the greater is its ability to be exchanged.

The sandy loam soil exhibited low CEC (5.6 meq/100 g) at its original pH of 4.7, indicating few exchangeable sites for adsorption of cations. Based on this CEC value, the soil can be classified as kaolinite (see Table VIII). Considering that the pH of the 30 percent subbituminous fly ash/soil leachate was slightly higher than that of the soil, the ability of the soil to exchange cations will remain almost the same. Consequently, the leachate quality of fly ash will not be greatly altered. The basis of this conclusion is that the smaller and higher-charged ions will adsorb onto the soil particles, quickly saturating the available sites. In contrast, the larger and lower-charged ions (most likely trace elements) tend to remain in solution.

Table VIII. Range of Cation Exchange
Capacity [11]

Clay Mineral	CEC (meq/100 g)
Kaolinite	3 - 15
Halloysite	5 - 40
Illite	10 - 40
Vermiculite	100 - 150
Montmorillonite	80 - 150

Table IX shows a comparison of the quantity leached of every element for leachates from subbituminous fly ash and 30 percent subbituminous fly ash/soil. Al, As, Cu, Pb, Mo and Zn were the only elements displaying a visible reduction in leachability from one leaching condition to another. Because the majority of these elements are present in trace amounts, the variation in the percentage of extractable amounts can be deceiving in the sense that absolute concentrations are low.

Table IX. Percentage Leached of Each Element from Subbituminous Fly Ash Relative to its Original Bulk Concentration tration under two Different Conditions.

Element	SFA Leachate (%)	30% SFA to Soil Leachate (%)
Aluminum	20.00	0.70
Calcium	> 100.00	> 100.00
Iron	0.04	0.04
Magnesium	15.00	> 100.00
Potassium	2.10	2.50
Silicon	0.10	0.80
Arsenic	15.70	3.20
Cadmium	12.50	56.30
Copper	0.10	0.00
Lead	9.10	2.40
Manganese	2.60	9.10
Molybdenum	37.60	7.20
Nickel	0.15	5.60
Phosphorus	0.50	0.55
Zinc	0.10	0.00

Water Quality Criteria

The National Interim Primary Drinking Water Standards (PDWS) specify maximum permissible concentrations of various constituents in domestic water supplies. Table X lists recommended criteria for some heavy metals along with the amount leached from subbituminous fly ash and 30 percent subbituminous fly ash/soil. Comparing the prescribed levels of these elements to the ones from the leachates, it is concluded that As, Cd, Fe, Pb and Mn exceed their recommended concentrations in drinking water.

Even though As, Cd, Fe, Pb and Mn exceeded their prescribed levels for drinking water, it would not be appropriate to state that the use of fly ash as a road base material is environmentally unsafe. The basis of such a statement is that the extraction procedure employed represents the worst case condition, since the material is overly exposed to the leaching solution. In the field, however, the low permeability of road bed retards

Table X. Leached Quantities of As, Cd, Cu, Fe, Pb, Mn and Zn from Sub-bituminous Fly Ash under two Different Conditions.

Element	SFA Leachate (mg/l)	30% SFA to Soil Leachate (mg/l)	PDWS Standards (mg/l)	RCRA Standards (mg/l)
Arsenic	1.20	0.24	0.05	5.00
Cadmium	0.01	0.43	0.01	1.00
Copper	0.02	0.00	1.00	--
Iron	0.64	0.55	0.30	--
Lead	0.54	0.14	0.05	5.00
Manganese	0.20	0.70	0.05	--
Zinc	0.01	0.00	5.00	--

the leaching of trace metals and hence migration of fly ash constituents to ground waters would be limited. In this situation, RCRA standards are the appropriate guideline, and the fly ash/soil mixture is well within permissible limits.

CONCLUSIONS

Based on the data derived from this study, the following conclusions can be drawn:

o   Due to the low CEC of the soil studied, the stabilization process was not very effective in altering the leachate quality of sub-bituminous fly ash.

o   Concentrations of As, Cd, Fe, Pb and Mn were above the recommended levels set by the Environmental Protection Agency in the National Interim Primary Drinking Water Standards, but well within RCRA standards.

o   Even though some elemental concentrations exceeded prescribed drinking water standards, it would not be justified to state that the use of fly ash on road bases is environmentally unsafe. In the field, the low permeability of the road bed will slow the leaching of elements. Furthermore, the migration of fly ash constituents to groundwater systems will be minimized because there is no direct access to groundwater, as is sometimes the case in ash disposal sites.

REFERENCES

[1] Babcock, A. W., and Faber, J. H., "Production, Availability and Utilization of Fly Ash," Ash Short Course, Louisiana State University, Baton Rouge, April 1979.

[2] Ledbetter, William B., "Is Lime in Fly Ash Available for Soil Stabilization?," Texas A&M University, College Station, April 1981.

[3] Thornton, Sam, "Use of Fly Ash for Soil Stabilization," Ash Short Course, Louisiana State University, Baton Rouge, April 1979.

[4] Dobie, Terrence R., Ng, Samuel Y., and Henning, Norman E., "A Laboratory Evaluation of Lignite Fly Ash as a Stabilization Additive for Soils and Aggregates," North Dakota State Highway Department, Bismark, January 1975.

[5] Ritter, Charles J., Bergman, Steven C., Cothern, Charles R., and Zamierowski, Edward E., "Comparison of Sample Preparation Techniques for Atomic Absorption Analysis of Sewage Sludge and Soil," University of Dayton, Dayton, Atomic Absorption Newsletter, Vol. 17, No. 4, July-August 1978.

[6] Soil Survey Staff, "Soil Investigation - Report No. 1," U.S. Department of Agriculture, Washington, D.C., 1972.

[7] Personal Communication, Environmental Protection Agency, Cincinnati, Ohio, 1984.

248

[8] Natusch, David F. S., and Taylor, David R., "Environmental Aspects of
     Western Coal - Chemical and Physical Characteristics of Coal Fly Ash,"
     Colorado State University, Fort Collins, November 1980.

[9] Linton, R. S., Williams, P., Evans, C. A., and Natusch, D. F. S.,
     "Determination of the Surface Predominance of Toxic Elements in
     Airborne Particles by Ion Microprobe Mass Spectrometry and Auger
     Electron Spectrometry," Analytical Chemistry, Vol. 49, No. 11,
     September 1977.

[10] Tittlebaum, M. E., "Investigation of Leachate Heavy Metal and Organic
     Carbon Content Stabilization through Leachate Recirculation," Ph.D.
     Dissertation, University of Louisville, Louisville, May 1979.

[11] Mitchell, J. K., Fundamentals of Soil Behavior, John Wiley and Sons,
     Inc., 1976.

TECHNICAL REVIEW OF THE ENERGY AUTHORITY COAL WASTE ARTIFICIAL REEF PROGRAM (C-WARP)

C.A. HORNIBROOK* AND J.H. PARKER**
*New York State Energy Research and Development Authority, 2 Rockefeller Plaza, Albany, New York 12223
**Science Applications, Inc., P.O. Box 509, New Port, Rhode Island 02840
(Formerly with Marine Sciences Research Center, State University of New York at Stony Brook, New York 11794)

(Received 4 January, 1985; accepted 11 March, 1985; Refereed)

ABSTRACT

The objective of the Coal Waste Artificial Reef Program (C-WARP) is to explore the technical feasibility and environmental effects of ocean disposal of stabilized fly ash and flue gas desulfurization scrubber sludge from power plants. This demonstration program has involved the construction of an artificial reef in the Atlantic Ocean off Long Island, New York using 15,000 blocks of these stabilized coal waste materials.

In determining the environmental acceptability of using stabilized coal waste blocks to build artificial reefs in the ocean, the two key considerations are whether the blocks will maintain their structural integrity, and whether the blocks will be a source of potentially toxic elements to affect the biological community. Analysis of the chemical composition of the blocks and an understanding of the processes occurring in them is necessary to address these questions. Therefore, this paper will focus on changes in chemical composition of blocks (e.g., Ca flux, Mg enrichment, and surface layer elemental concentrations) and compatibility of the blocks in terms of the physical and biological data collected during three years of monitoring.

INTRODUCTION

In 1978 the New York State Energy Research and Development Authority, the U.S. Department of Energy, the Electric Power Research Institute, the Environmental Protection Agency and the Power Authority of the State of New York cosponsored the Coal Waste Block Artificial Reef Program. The research has been carried out by the Marine Science Research Center (MSRC) at the State University of New York at Stony Brook. The project was initiated in response to the growing need for disposal alternatives for coal combustion waste products. Electric utilities were conducting feasibility studies on converting oil fired power plants to coal. Many new facilities were also designed for coal. The waste disposal problem for New York State alone was projected to grow from 800,000 tons of ash in 1978 to 10.8 million tons of ash and flue gas desulfurization (FGD) sludge by 1991. (FGD sludge potentially doubles the amount of waste produced at each coal fired facility.)

Hence, the objective of the Coal Waste Artificial Reef Program (C-WARP) has been to explore the technical feasibility and environmental effects of the ocean disposal of stabilized fly ash and FGD sludge from power plants. The research has progressed from laboratory studies to a demonstration reef (September 1980) off Fire Island in the Atlantic Ocean. The reef consists of 500 tons of coal combustion wastes processed into 15,000 stabilized blocks. The following discussion addresses specific research areas covered during the conduct of this multiyear program. The engineering aspects of block production, handling, and ocean placement; ecological implications of block interactions with the aquatic environment; and the physical and chemical integrity of the blocks are discussed.

BLOCKS AND REEF BUILDING

Building the demonstration reef required thousands of blocks. Existing methods had to be adopted in order to quickly produce cost-effective blocks. In collaboration with the Besser Company, Alpena, Michigan, techniques were developed to process power plant wastes with machinery and equipment presently employed by the concrete industry. More than 20 combinations of fly ash/FGD sludge mixtures were tried under various test conditions. The blocks from these mixes were tested for handling suitability (by automated equipment), hardening and curing qualities, strength[1], etc. Additives tested during this phase were: hydrated lime (6% by weight), quick lime (6% by weight), and Type II portland cement (3 to 6% by weight). Cured blocks were tested for impact resistance and compression strength.

The techniques developed at the Besser Company were successfully transferred to a commercial concrete block factory and 500 tons of power plant FGD scrubber sludge and fly ash were processed to manufacture 15,000 solid blocks of stabilized material 20 x 20 x 40 cm (8 x 8 x 16 inches). The blocks were formed out of two coal waste mixes from different sources with respective fly ash-to-scrubber sludge ratios of 3:1 and 1:1. During the block making process, 6% hydrated lime and 3% portland cement were added to the FGD sludge and fly ash and thoroughly mixed. This mixture was run into the hopper of a block machine (Fig. 1) and then fed into steel molds. The molded blocks were then palletized.

These factory-made blocks were then cured in on-site kilns to achieve compressive strengths of about 500 psi, which allowed for stacking and easy transport. The stabilized blocks were trucked to a terminal on the Hudson River estuary and loaded on a barge fitted with bottom-opening doors (Fig. 2). On 12 September 1980 the barge released the blocks at the project site (Fig. 3), which is two miles south of Long Island in the Atlantic Ocean at a depth of approximately 20 meters (68 feet).

The reef forms one continuous structure approximately 77 m (250 ft.) long, 14-18 m (45-70 ft.) wide, and 1-2 m (3-6 ft.) high. Repeated oceanographic surveys were made at and near the project site prior to reef construction in order to characterize the physical and chemical properties of the water column, nutrient concentrations, suspended particulates, and current speed and direction, etc.

---

[1] The development of early block strength relied heavily upon the curing schedule.

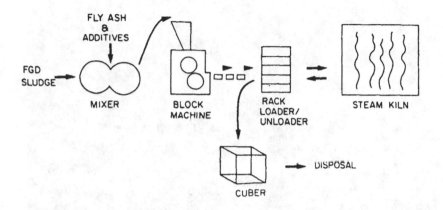

Figure 1. Simplified schematic of block processing. (1)

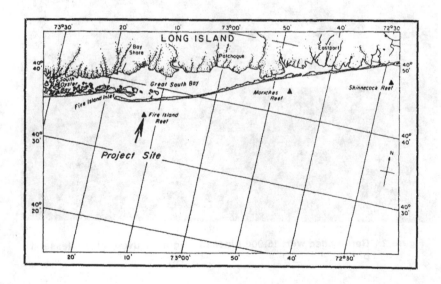

Figure 3. C—WARP project site. (1)

Figure 2.   Barge filled with 15,000 blocks being maneuvered for release at
project site.   (2)

BLOCK CHEMISTRY

Knowledge of the composition, mineralogy, and seawater dissolution of stabilized coal waste blocks and their components is fundamental to understanding and predicting behavior of these blocks in the marine environment. Throughout the duration of C-WARP, a wide variety of analyses have been performed on the test blocks, including their chemical composition, mineralogical changes, and ultimately their compressive strength (see Physical Properties).

Chemical Composition

Coal waste materials to be used for the reef blocks were obtained from the Conesville power station of the Columbus and Southern Ohio Electric Company, and the Petersburg station of the Indianapolis Power and Light Company (IPALCO). Table 1 gives the chemical compositions of fly ashes and FGD sludges from which the blocks were made and the composition of the finished blocks. In all the materials, calcium (contributed by the FGD sludge) is the major component, comprising as much as 30% of the block material. The fly ashes from the different sources vary in composition, but all contribute to the metallic components Ae, Si, Fe, and the trace elements. The fly ash-to-FGD sludge ratios of the blocks are, therefore, reflected in their chemical compositions.

Method Development

Because the coal wastes are composed of a variety of mineral phases, a proper digestion technique was required to accurately determine the elemental composition by atomic absorption spectrophotometry. Comparative experiments were conducted on National Bureau of Standards standard materials for several digestion methods including HNO_3, HCl, HF/H_3BO_3, and $HF/H_3BO_3/HNO_3$.

From the experimental results the HF/H_3BO_3 digestion was considered acceptable for the analysis of coal waste materials. For the analysis of As, HCl digestion was found to be most accurate and precise, yielding 100% recovery. Other methods of chemical analysis were used to confirm the digestion and atomic absorption spectrophotometric techniques (e.g., x-ray fluorescence, neutron activation, and microprobe analysis).

Because of the inherent non-homogeneity of the stabilized coal waste blocks, a procedure was developed to obtain representative samples for chemical analysis. The procedure was to collect 50 grams of crushed block material while the block was wet to prevent migration of elements and salt deposits. Samples were taken from the surface and core regions of the block to reflect possible compositional changes (Fig. 4).

TABLE 1

CHEMICAL COMPOSITION OF FIXATED COAL WASTE BLOCKS

Sample Analysis	Al %	As µg/g	Ba µg/g	Ca %	Cd µg/g	Co µg/g	Cr µg/g	Cu µg/g	Fe %	K %
Blocks										
Conesville 2.0:1* Surface	7.70 ±0.15	63.8 ±1.3	652 ±71	10.2 ±0.4	0.74 ±0.03	26.0 ±0.5	90.0 ±3.2	60.0 ±1.2	15.3 ±0.3	1.17 ±0.03
Conesville 3.0:1 Core	6.26 ±0.41	67.6 ±1.4	676 ±67	10.0 ±0.2	0.78 ±0.03	27.3 ±0.6	95.6 ±2.9	63.4 ±1.3	15.3 ±0.3	1.06 ±0.02
IPALCO 0.8:1 Surface	5.40 ±0.16	47.3 ±1.3	353 ±83	19.1 ±0.6	1.48 ±0.04	26.2 ±0.5	70.8 ±3.6	67.1 ±1.4	6.34 ±0.18	1.23 ±0.02
IPALCO 0.8:1 Core	5.47 ±0.11	44.2 ±2.2	340 ±110	18.6 ±0.9	1.31 ±0.04	25.9 ±0.5	70.9 ±3.6	64.5 ±1.3	6.05 ±0.11	1.28 ±0.05
FGD Sludges										
Conesville	0.31 ±0.01	0.7 ±1.29	51.5 ±44.8	25.5 ±2.83	0.14 ±0.04	0.70 ±0.39	9.67 ±3.47	3.71 ±1.05	0.21 ±0.01	0.04 ±0.00
IPALCO	0.23 ±0.01	3.12 ±1.38	0.0 ±20.8	27.8 ±1.91	0.46 ±0.04	1.51 ±0.42	9.22 ±3.71	4.31 ±1.12	0.38 ±0.01	0.00 ±0.21
Fly Ashes										
Conesville	11.32 ±0.23	86.9 ±1.74	773 ±92.0	1.56 ±0.05	0.96 ±0.04	33.3 ±0.68	114.0 ±3.16	80.7 ±1.62	17.1 ±0.34	1.75 ±0.02
IPALCO	10.53 ±0.21	76.5 ±3.84	379 ±22.3	2.64 ±0.05	3.95 ±0.08	53.9 ±1.09	149.0 ±3.52	122.0 ±2.46	19.3 ±0.38	1.77 ±0.02

The ± denotes error in reported value calculated from sampling, analysis, and standards errors.
*Fly Ash/FGD Sludge

TABLE 1
(Continued)

Sample Analysis	Li μg/g	Mg %	Mn μg/g	Na %	Ni μg/g	Pb μg/g	Si %	Sr μg/g	Ti %	Zn μg/g
Blocks										
Conesville 3.0:1* Surface	132 ±3	0.66 ±0.01	197 ±10	0.20 ±0.00	81.0 ±1.6	31.9 ±1.0	14.6 ±0.1	500 ±25	0.42 ±0.00	156 ±3
Conesville 3.0:1 Core	122 ±2	0.72 ±0.02	210 ±4	0.17 ±0.00	84.3 ±1.7	33.0 ±0.9	15.6 ±0.3	487 ±24	0.42 ±0.02	160 ±5
IPALCO 0.8:1 Surface	59.7 ±1.19	0.67 ±0.00	167 ±7	0.17 ±0.00	85.3 ±1.7	26.7 ±1.1	11.3 ±0.7	396 ±16	0.28 ±0.00	199 ±4
IPALCO 0.8:1 Core	61.0 ±1.8	0.67 ±0.02	170 ±8	0.17 ±0.00	87.6 ±1.7	25.1 ±1.1	11.8 ±0.5	388 ±31	0.28 ±0.00	222 ±4
FGD Sludges										
Conesville	2.51 ±0.48	0.63 ±0.02	22.1 ±5.07	0.24 ±0.01	4.19 ±1.62	0.53 ±0.04	2.95 ±0.10	230 ±6.93	0.02 ±0.00	16.5 ±2.05
IPALCO	3.42 ±0.52	0.05 ±0.00	25.4 ±5.42	0.01 ±0.00	4.08 ±1.73	1.97 ±1.12	0.50 ±0.02	230 ±6.44	0.01 ±0.00	23.7 ±2.19
Fly Ashes										
Conesville	169 ±5.08	0.62 ±0.02	232 ±4.62	0.23 ±0.01	91.4 ±1.84	66.9 ±1.33	21.0 ±0.44	572 ±17.5	0.59 ±0.02	194 ±3.88
IPALCO	134 ±2.68	0.52 ±0.02	290 ±23.1	0.35 ±0.01	166 ±3.33	113 ±2.25	19.6 ±0.39	382 ±7.66	0.55 ±0.03	506 ±21.0

The ± denotes error in reported value calculated from sampling, analysis, and standards errors.
*Fly Ash/FGD Sludge

Changes in Block Composition

Measurements of the concentrations of the elements Ca, Mg, Na, K, Si, Al, Zn, Mn, Cd, Pb, As, Cu, and Ni were made in both the core and the surface (1 cm) of Conesville and IPALCO blocks returned to the laboratory from the artificial reef. These data present the average changes in concentration of these elements after being exposed in the sea for 132 weeks and can be found in P.M.J. Woodhead, et al. (1984). Only six elements (Ca, Mg, Na, K, Si, and Al) exhibited any significant changes in concentration over the exposure period. Figures 5 to 9 indicate changes in the surface concentrations relative to those of the core sections of the blocks. Because of the slow diffusion rates for the elements in the interior areas of the blocks, the core concentrations have not changed significantly from the initial values obtained before exposure to seawater.

The most dramatic changes occurred in the Ca and Mg concentrations. As calcium (the major component of the block) diffuses from the blocks, an ion exchange process creates an enrichment of Mg (from the seawater). Over the 132 week period, the average mole Ca loss/mole Mg gain ratio has been approximately 1.5. The lack of deterioration (see Physical Properties) tends to confirm that the loss of Ca does not affect the block's structural integrity. The increases in Na concentration throughout the entire block indicate saturation with seawater. Yet the concentrations continued to increase beyond those expected from saturation, indicating the possibility of another active chemical process taking place. The elements Al and Si, contributed by the fly ash, appear to increase in concentration relative to the loss of Ca in the blocks. The relative increases in Al and Si, which could be interpreted as an increase in fly ash concentrations, are explained below.

Fly Ash Concentrations

Fly ash appears to be an inert component of the coal waste blocks with respect to seawater dissolution. Therefore, the concentrations of elements associated with it do not experience significant reduction. Because fly ash comprises up to 50% of some blocks, relative increases in fly ash concentrations occur because of the loss of Ca and other FGD components from the blocks.

---

*Note: The fly ash-to-scrubber sludge ratio for Conesville is 3:0:1. Conesville-Y is 2.9:1, IPALCO is 1.5:1, and IPALCO Y is .8:1. Due to the similar reaction in all blocks analyzed all data are presented.

BLOCK SAMPLING

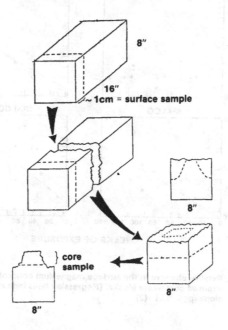

8"

16"
~1cm = surface sample

8"

core
sample

8"

8"

Figure 4. Sampling procedure for chemical analyses. (3)

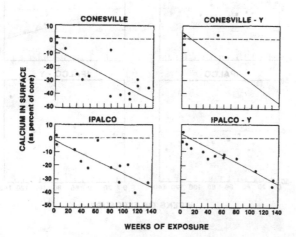

CALCIUM IN SURFACE
(as percent of core)

CONESVILLE

CONESVILLE - Y

IPALCO

IPALCO - Y

WEEKS OF EXPOSURE

Figure 5. Relative changes in the surface calcium concentrations of exposed
coal waste blocks. [Regression lines indicate significant slope
($p \leq 0.05$).] (3)

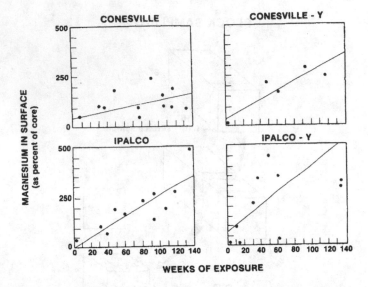

Figure 6. Relative changes in the surface magnesium concentrations of exposed coal waste blocks. [Regression lines indicate significant slope (p≤0.05).] (3)

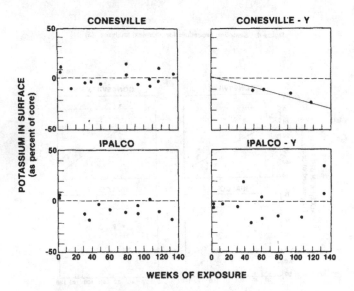

Figure 7. Relative changes in the surface potassium concentrations of exposed coal waste blocks. [Regression lines indicate significant slope (p≤0.05).] (3)

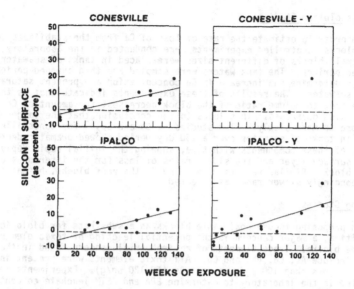

Figure 8. Relative changes in the surface silicon concentrations of exposed coal waste blocks. [Regression lines indicate significant slope (p≤0.05).] (3)

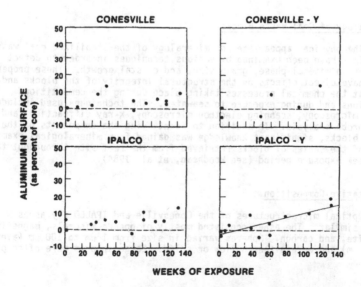

Figure 9. Relative changes in the surface aluminum concentrations of exposed coal waste blocks. [Regression lines indicate significant slope (p≤0.05).] (3)

Tank Dissolution Studies

In order to estimate the rate of loss of Ca from the stabilized coal waste blocks, controlled experiments were conducted in the laboratory, where small blocks of different sizes were placed in tanks of seawater for extended periods. The tank waters were sampled and then changed periodically to determine the increase in Ca concentration and prevent saturation of the seawater. The results of these experiments indicate that as the surface area to volume ratio of the block decreases, the amount of Ca leached also decreases. This leads to the conclusion that the larger block will lose Ca more slowly. Flux studies of Ca and SO_4^{2-} for Conesville blocks in tanks of seawater over a 150 day period showed dramatic decreases in flux of these components with time, showing the initial rapid depletion in the surface layer and the slower rates of loss for the interior sections of the block. Similar results were found in the reef blocks, but at a correspondingly slower rate, as expected.

Leaching Studies

In proposing to use coal waste blocks as a substrate for biological communities, a major concern is the potential leaching of toxic elements from the blocks. Trace elements of environmental concern found in the blocks include As, Cd, Pb, and Cu. All toxic elements were present in the blocks at less than 100 μg/g, except As (60-120 μg/g). Experiments were conducted in the laboratory to determine EPA and ASTM leachate concentrations and compare them to EPA drinking water quality criteria. Examination of leachate data, Table 2, reveals that no elemental concentration exceeded the EPA criteria, with the exception of cobalt and manganese. Comparison of the EPA and ASTM procedures indicates that EPA leachate concentrations can be up to 10 times higher than the ASTM values for many elements.

Mineralogy of Coal Waste Blocks

The physical appearance and mineralogy of the stabilized coal waste materials have been examined by various techniques in order to detect changes in mineral phase, grain size, and crystal growth. These properties will have direct effects on the structural integrity of the blocks and suggest the chemical processes taking place during the cementitious reactions and during exposure to seawater. The techniques used include light microscopy, scanning electron microscopy, x-ray diffraction, and microprobe analysis. In addition to identifying the components of the coal waste blocks, significant knowledge was gained from mineralogical examination of the series of blocks retrieved from the reef site throughout the 132-week exposure period (see Woodhead, et al. 1984).

Qualitative Composition

Optical microstructures of the Conesville and IPALCO fly ashes were quite similar. The ashes consisted mostly of opaque minerals, magnetite, hematite, and carbon. Spheres varied in size from 1 μm to 100 μm (average of 10 m) and were either hollow or solid. Large spheres were often pitted

TABLE 2

CHEMICAL COMPOSITION OF EPA EP LEACHATE AND DRINKING WATER QUALITY CRITERIA (2)

Sample	Al (ppm)	As (ppb)	Ba (ppm)	Cd (ppb)	Co (ppb)	Cr (ppb)	Cu (ppb)	Fe (ppm)	K (ppm)
IPALCO 0.8:1	3.99 ±0.36	3.87 ±0.63	1.78 ±0.94	8.03 ±0.26	19.4 ±1.74	11.7 ±1.78	43.8 3.94	0.25 ±0.02	73.8 ±6.65
Conesville 3.0:1	2.84 ±0.35	14.7 ±1.32	0.32 ±0.32	4.56 0.26	29.2 ±2.63	26.0 ±2.34	22.4 ±2.02	0.44 ±0.04	55.9 ±5.03
Drinking Water MCL's		50.0		1.0	10.0	1000	50.0	1000.0	0.3

Sample	Li (ppm)	Mg (ppm)	Mn (ppm)	Na (ppm)	Ni (ppb)	Pb (ppb)	Si (ppm)	Sr (ppm)	Ti (ppm)	Zn (ppm)
IPALCO 0.8:1	0.07 ±0.01	18.6 ±1.67	0.47 ±0.04	14.0 ±2.13	65.5 ±5.89	1.39 ±0.22	40.1 ±3.61	3.62 ±0.33	0.00 ±0.05	0.23 ±0.01
Conesville	0.22 ±0.02	46.2 ±4.16	0.52 ±0.05	14.3 ±2.13	108 ±9.71	0.97 0.22	45.5 ±0.92	4.13 ±0.37	0.05 ±0.05	0.10 ±0.01
Drinking Water MCL's			0.05			50.0				

and contained aluminates and silicates (mullite) and amorphous material.
Quartz and calcite were present in small amounts.

Conesville FGD sludge consisted mainly of calcium sulfite hemihydrate
($CaSO_3.\frac{1}{2}H_2O$), while IPALCO scrubber sludge consisted of about 70% hemihydrate and 30% gypsum ($CaSO_4$). The phases took the form of small plates
ranging in width from 1 to 7 microns.

Evidence of Reactions

Prior to reef construction, test blocks were placed in seawater in the
laboratory. Although fly ash is fairly inert, the FGD sludge component of
the blocks contains elements that are more soluble and these elements can
be exchanged with ions in the seawater. Blocks exposed to seawater
experienced a loss of portlandite ($Ca(OH)_2$) and a gain of brucite
($Mg(OH)_2$), $MgSO_4$, and ettringite. Prior to exposure, tubular crystals of
calcium aluminosilicates are observed along with the amorphous material.
After exposure, the plate-like forms of scrubber sludge include gypsum and
$CaSO_3.\frac{1}{2}H_2O$. Crystals of ettringite are seen growing out of gypsum
crystals. Macroscopic examination of blocks retrieved from the reef site
revealed a dark region, or rim, at the surface of the block, indicating the
reaction with seawater. Data collected on the change in rim width throughout the exposure period are reported in detail by P.M.J. Woodhead, et al.
(1984).

Recrystallization

Scanning electron micrographs were examined for changes in grain size
in a Conesville block after one-year exposure in the sea. No evidence of
recrystallization was detected. Volumes of crystalline and non-crystalline
matrices determined for blocks from the reef site for one-year reveal
little significant change. Because of the inherent block nonhomogeneities,
these qualitative examinations may not detect evidence of
recrystallization. However, the occurrence of the dark rims near the
surface of the blocks and the changes in block composition measured in the
elemental analysis suggest that some alteration processes do occur.

PHYSICAL PROPERTIES

The C-WARP program has shown that the porosity and effective elastic
modulus of the consolidated coal waste material provide the best measure of
its integrity in the marine environment. Over the three years of sampling,
elastic modulus was monitored by measuring the velocity of sound propaga-
tion in situ at the reef and on blocks recovered from the reef for
evaluation in the laboratory. The physical testing program concentrated on
the measured and derived physical properties of ultrasonic velocity,
ultrasonic attenuation, effective elastic modulus, compressive strength,
density, and porosity.

A direct comparison of elastic modulus to compressive strength values
can be made for the coal waste blocks. The elastic modulus varied in
parallel with the compressive strength throughout the study period. The

compressive strengths appear to have stabilized at a value of 6.5 MPa over
the last year of testing.

Though the majority of results have been positive, there have been
some that indicate a potential loss in block integrity. This can only be
borne out through future periodic checks of the blocks' physical
characteristics.

HABITATION OF REEF BY FISHES

It should be noted that the Coal Waste Artificial Reef was constructed
some 700 feet from an existing Rubble Reef. The Rubble Reef was built 20
years earlier by the New York State Department of Environmental Conservation and consists mainly of construction rubble. An outcropping of the
Rubble Reef, similar in size to the Coal Waste Reef, was used as a control
by which to measure the relative success of the progression of habitation
that occurred at the Coal Waste Reef.

Five weeks after placement on the seabed, cunner were the initial
inhabitants and remained the most numerous fish. Cunner is also the most
abundant species on the nearby Rubble Reef. Other species found at the
Coal Waste Reef during the first year's surveys were black sea bass,
blackfish, conger eel, winter flounder, summer flounder, ocean pout, sea
raven, and longhorn sculpin. Adult rock crabs were early migrants to the
reef and toward the end of the first year, during the summer of 1981,
juvenile lobsters, 10-25 cm in total length, began to appear in increasing
numbers.

The fish populations of both reefs were sampled at the same times
using the same techniques over a three year period. Because the areas of
the two reefs were approximately equal, it was possible to make meaningful
comparisons between the fish populations.

The average trap catches, in catch per unit effort, of fish from both
reefs showed marked change in relation to the seasons and more clearly to
the annual cycle of sea temperatures. Though trap catches were smaller on
the Coal Waste Reef than on the Rubble Reef, they were qualitatively the
same (i.e., the same fish species became year-round residents of the new
reef, as were residents on the Rubble Reef, and seasonally transient
species appeared and disappeared from the two reefs at precisely the same
time. However, the catch differential between the reefs was closely
maintained in 1980-81. By 1982, however, the fish populations on both
reefs were of the same size. The cunner populations were numerically
dominant on both the Coal Waste Reef and the Rubble Reef and showed little
change between one reef and the other (Table 3). This allowed measurement
of a number of fundamental population parameters for comparison between the
two resident cunner populations. Cunner on the Coal Waste Reef were of
similar size distributions with fish on the Rubble Reef, although there
were more fish in the larger size classes on the Coal Waste Reef in each
year from 1981 to 1983. The sex ratios (which are naturally skewed toward
females in cunner populations) were different during the first year for the
new reef, but thereafter there was no difference from the Rubble Reef
fishes.

TABLE 3
ESTIMATES OF POPULATION SIZE AND DENSITY FROM RECAPTURED TAGGED CUNNER

	1981	1982	1983
Population Size			
Coal Waste Reef (1,230m²)	3,421*	10,171	11,000
Fire Island Reef (1,215m²)	10,107	9,588	12,936
Population Density			
Coal Waste Reef	2.7/m²*	8.3/m²	8.9/m²
Fire Island Reef	8.3/m²	7.9/m²	10.6/m²

*Significantly smaller (p 0.01) than the population on the Rubble Reef in 1981.

Population size and density were estimated by fish tag-and-recapture experiments for the cunner resident on the Coal Waste Reef and compared with the population on the Rubble Reef. Although the Coal Waste population was significantly smaller (p 0.01) during the first year (1980-81), the populations were of the same size and density in 1982 and 1983. Estimates of survival and mortality rates by two different methods, made from samples of fish caught on each reef, showed no differences between these parameters for the two populations. These are conventional indices for expressing survival in fish populations and the relative survival rates were the factors that were not found to be different.

TABLE 4
ESTIMATES OF SURVIVAL RATE (S) AND MORTALITY RATE (Z)
FOR CUNNER POPULATION

Population	Robson & Chapman[1]		Ricker[2]		
	S	Z	S	Z	n
Coal Waste Reef[3]					
1981	0.56	0.58	0.76	0.27	414
1981 + 1982	0.55	0.60	0.77	0.26	538
Fire Island Reef[4]					
1981	0.62	0.48	0.74	0.30	793
1981 + 1982	0.56	0.58	0.75	0.29	940

[1]Robson & Chapman, 1960.   [3]Estimated ages 4-9 inclusive.
[2]Ricker, 1975.   [4]Estimated ages 5-10 inclusive.
n number of fish caught

Our conclusion at this time is that the coal waste is a suitable habitat for residence by a community of fish typical of artificial fishing reefs in the Northeast, and it appears that the fish inhabiting the reef are not anomalous in any way.

SUBSTRATE SUITABILITY

The stabilized coal wastes were tested for their suitability as substrates for the settlement and growth of an encrusting, sessile community of invertebrates. Racks containing sets of test bricks just above the seabed allowed settlement and attachment by different types of organisms on the bricks. Larval settlement was studied by setting out clean test bricks for relatively short exposures of one to two months to determine which species settled and attached to the bricks at different times of the year. To study colonization and community development, test bricks of the same waste material in the sea for longer periods of up to two years were sampled to measure the growth and differentiation of the communities encrusting on them. As you will recall, the reef is in 68 feet of water, and, therefore, below the photic zone. For this reason, all growth is animal life and not plant life. Therefore, during the three years fall 1980 to spring 1983, the fauna growing on the coal waste reef blocks and on concrete blocks were very similar. (Note that at the time of reef construction, concrete blocks were mixed in with the coal waste blocks to act as a control material on the reef itself.) After the first year in the sea, growths on the Conesville, IPALCO and concrete were about the same (Fig. 10). In addition, the same patterns of seasonal increase and decrease in percent cover by communities were followed on all of the block types at the same time. The conclusion here is that the blocks provide a suitable substrate for organisms to settle and flourish and the organisms that settled and colonized are typical of many hard bottom epifaunal communities in coastal waters of the Northeast seaboard.

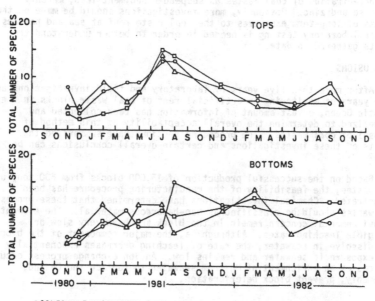

○ CONESVILLE; □ CONCRETE, △ IPALCO.

Figure 10. Number of species counted on the test bricks. (3)

TOXIC POTENTIALS

The potential toxicity of the coal waste materials was tested via laboratory analysis and quantitative characterization of the wastes to determine trace element content and leaching behavior in different aquatic systems. The results of U.S. EPA "E.P. tests" indicated that leachates from the wastes would not be classified as hazardous under the Resource Conservation and Recovery Act of 1976. Short-term laboratory assays for toxicity and longer experiments to measure elemental uptake by test organisms in aquaria were generally negative (e.g., inhibitory effects were not demonstrated).

Mussels held in cages for many months at the reef did not acquire elevated tissue concentrations of trace elements nor did the fish (cunner) which is resident at the reef site (Table 5).

The findings from analyses for uptake of chemical components of the blocks (including elements of environmental concern) by the reef-associated fauna in the sea are persuasive. Exposures to the coal waste materials extended over many months without evidence for tissue accumulation of trace elements. The results of a few of the laboratory experiments did indicate some effect due to the coal wastes. In general the elements of concern were arsenic, iron, and nickel which accumulated in the tissues of mussels that were exposed for long periods to high concentrations of suspended coal wastes in laboratory fish tanks. It is interesting to note that several heavy metals showed no accumulation in the tissues of mussels exposed to any concentration of coal wastes as suspended sediment lead, mercury, manganese and zinc. Obviously, more investigations should be made on the effects of long-term exposures to the coal waste reef at sea and perhaps further laboratory testing is needed in order to better understand the results gathered to date.

CONCLUSIONS

After more than five years of laboratory and field investigations and three years of monitoring an artificial reef of coal waste blocks in the Atlantic Ocean, a vast amount of information has been collected and synthesized to determine the overall acceptability of this method of disposal of coal wastes. The preceding sections have summarized the results of these investigations and certain overall conclusions can be made.

Based on the successful production of 15,000 blocks from 500 tons of coal wastes, the feasibility of the manufacturing procedure has been demonstrated. Chemical investigations have determined that these processed coal wastes should be classified as non-hazardous material. The elements of environmental concern remain in the block and have very slow or negligible leaching rates. Although Ca, the major component of the blocks, does dissolve in seawater, the rate of leaching decreases exponentially upon exposure to seawater and remains low. An ion exchange process occurs that causes an enrichment of Mg from seawater to replace the Ca and, presumably, prevent block deterioration.

TABLE 5

TRACE METAL CONCENTRATION IN TISSUES OF MUSSELS EXPOSED IN THE SEA FOR 5 MONTHS
AND FOR 9 MONTHS AT THE COAL WASTE BLOCK REEF AND AT A CONTROL SITE ON THE SEABED

Treatment Sample Date	Elemental concentration (μg/g dry weight)							
	As	Cd	Cu	Fe	Mn	Ni	Pb	Zn
Initial Concentration 2/81	6.8±0.4	1.2±1.0	11.9±1.5	56.6±11.0	11.5±2.6	1.2±0.3	1.5±0.2	76.8±13.9
Batch 1								
Controls 7/81	3.4±1.1	1.0±0.1	5.6±1.1	54.3±9.5	9.2±2.6	1.1±0.3	1.0±0.3	72.3
Waste Reef 7/81	3.6±1.6	0.7±0.1	6.1±1.7	62.0±16.2	9.6±2.7	1.4±0.7	1.1±0.5	73.5
Batch 1								
Controls 11/81	5.7±2.2	1.3±0.5	8.8±3.1	60.1±24.4	-	1.3±0.6	2.2±1.7	160.5±35.5
Waste Reef 11/81	5.1±1.2	1.3±0.2	6.2±0.6	44.7±8.0	-	1.1±0.3	1.3±0.6	111.3±27.2
Batch 2								
Controls 11/81	7.7±1.9	1.6±1.7	6.7±0.8	73.8±24.1	10.3±3.8	2.4±0.5	2.3±0.7	185.7±81.3
Waste Reef 11.81	6.8±2.5	1.4±0.3	6.9±1.0	52.9±8.2	7.2±4.5	1.9±0.8	1.5±0.4	135.5±41.8

The behavior of the physical properties of the coal waste blocks upon exposure to seawater demonstrates stable or increasing trends in compressive strengths, density, and sound velocity. Variations in the porosity of the blocks reflect the different chemical changes occurring and possible related volume changes. No evidence of significant deterioration was detected. Increases in biological colonization on the surfaces of the blocks may help to prevent any erosive factors and reduce diffusion.

Diverse biological investigations have included experimental laboratory studies, but most of the investigations were carried out in the sea at the reef site. Many quantitative comparisons were made between organisms exposed to coal waste materials or derivatives and control organisms; in virtually all instances, the coal waste materials had no detectable effects on the test organisms. However, at very high concentrations of waste materials, some inhibitory or negative results were occasionally obtained and warrant further investigation. The reef appears to have no effects upon the organisms growing on the blocks or using the reef as a habitat.

Periodic sampling will be necessary to confirm that the coal waste reef can maintain its integrity. However, the demonstration reef to date appears to be without adverse effects in the sea. Fixation and consolidation of coal wastes to form reef blocks may provide an acceptable alternative to problems of disposal of coal combustion wastes from power plants. That solution would also carry benefits for the marine recreational fishing community through reef construction and marine habitat improvement.

BIBLIOGRAPHY

1. Parker, J. H., Woodhead, P. M. J., Duedall, I. W., Carleton, H. R. "Coal Waste Artifical Reef Program, Phase 4A," EPRI CS-2574, (1982).

2. Woodhead, P. M. J., Parker, J. H., Duedall, I. W. Marine Fisheries Review, 44, (1982).

3. Woodhead, P. M. J., Parker, J. H., Carleton, H. R., Duedall, I. W. "Coal Waste Artificial Peef Program, Overall Summary Report" (in publication).

4. Woodhead, P. M. J., Parker, J. H., Carleton, H. R., Duedall, I. W. "Coal Waste Artificial Reef Program, Phase 4B", EPRI CS-3726 (1984).

Author Index

Subject Index

Printed in the United States
By Bookmasters